Innovations in Earth Sciences

The sciences move fast—

and ABC-CLIO's Innovations in Science series can help you keep pace. Each title provides an overview of the events, scientists, and innovations that have shaped the development of a particular field of science during the past hundred years. Well suited to student research, or just for satisfying the curious, these accessible handbooks for the nonspecialist highlight the scientific breakthroughs of the twentieth century and the prospects for the twenty-first.

Titles in This Series

Innovations in Astronomy
Innovations in Biology
Innovations in Earth Sciences

INNOVATIONS IN SCIENCE

Innovations in Earth Sciences

Overview by Edward D. Young

ABC-CLIO

Santa Barbara, California
Denver, Colorado
Oxford, England

Library of Congress Cataloging-in-Publication Data

Innovations in earth sciences.
p. cm. — (Innovations in science)
Includes bibliographical references.
1. Earth sciences. I. Series
QE26.2.I56 1999 550—dc21 99-27925

ISBN 1-57607-115-4 (alk. paper)

05 04 03 02 01 00 99 10 9 8 7 6 5 4 3 2 1 (cloth)

ABC-CLIO, Inc.
130 Cremona Drive, P.O. Box 1911
Santa Barbara, California 93116-1911

This book is printed on acid-free paper ∞.

Contents

Preface

Scientists engaged in the study of Earth and the other rocky bodies of the solar system are troubled by what can only be described as a crisis of identity. The crisis is deep. What was a geologist 20 years ago is now an earth scientist. But is a chemist studying the oceans an earth scientist? Academic departments worldwide are vexed. What are they to call themselves? Should they be known as the Department of Geology, the Department of Earth Sciences, the Department of Earth and Environmental Sciences, the Department of Geological and Geophysical Sciences, or the Department of Earth and Planetary Sciences? In some quarters it is even suggested that the activities of scientists studying Earth can no longer be described as belonging to a single discipline, and that just as it is rare to find the life sciences under a single roof in most universities today, so too will go the earth sciences.

The present plight of earth scientists can be traced to the evolution of their discipline from the "natural history" of the eighteenth and nineteenth centuries to one in which revolutions in the physical and life sciences of the twentieth century are applied to the study of Earth and the solar system. The *American Journal of Science,* despite its all-inclusive name, is devoted to the publication of original contributions in the geological sciences and was so named in the nineteenth century to underscore this transition. Now, at the end of the twentieth century journal names again reflect a sea change in the earth sciences. For example, members of the Mineralogical Society of America debate whether or not to change the name of their journal, the *American Mineralogist,* to *The Journal of Earth and Planetary Materials* because it is thought that the former connotes an outdated and limited scope while the latter better reflects the flow of ideas that takes place between, say, mineralogists (or are they mineral physicists?) and material scientists studying the crystalline structures of high temperature superconductors.

Examples of synergism among earth science and the other core sciences—chemistry, physics, and biology—abound in the twentieth century. Indeed, the story of how our understanding of planet Earth has grown over the past century is inextricably linked with the developments in science in general.

Arrangement of This Volume

Part I of *Innovations in Earth Sciences* comprises six chapters. The first is an introductory essay that presents an overview of the earth sciences in the twentieth century and highlights the most significant developments.

Chapter 2 traces events in the earth sciences in chronology form, while readers can turn to chapter 3 for a selection of sketches of the most famous and influential individuals who have made contributions to the earth sciences field in the twentieth century.

Chapters 4, 5, and 6 serve as springboards for further information: they offer an annotated directory of earth-sciences-related organizations, a selected further reading list, and a sampling of Web sites, respectively.

Part II of this volume, Dictionary of Terms and Concepts, defines nearly 800 key terms and concepts used in the earth sciences, some of them illustrated with diagrams. This section can also be a useful adjunct to readers' gaining a thorough understanding of discussion in other parts of this handbook. Finally, at the back of the book readers will find several statistical charts as well as a general subject index.

Innovations in Earth Sciences

Part I

1 Overview

The collective term *earth sciences* is sometimes incorrectly regarded as a modern substitute for geology. When scientists trained in any scientific discipline turn their attention to the study of Earth, they are engaged in the earth sciences. The field is therefore unusually broad, encompassing studies of Earth's atmosphere, oceans, crust, and the structure and composition of the planet at great depth. It even extends to the study of the other bodies of the solar system, including the so-called terrestrial planets. Depending upon their subject of study, biologists, chemists, physicists, mathematicians, and material scientists can all be counted among earth scientists. Clearly, geology is only one part of this burgeoning field.

The diversity of the earth sciences is a consequence of the scientific revolutions that transformed the "natural history" of the eighteenth and nineteenth centuries to the modern physical and life sciences of the twentieth century. These revolutions facilitated the transition from mere description of the earth to a more rigorous, "scientific" approach to understanding the interactions among our planet's varied constituents and its evolution as a whole. Today, rather than simply describing layered rocks comprising mountain ranges as they might have done in the nineteenth century, geologists are relating what they have learned about the formation of those mountains to the chemistry of the oceans and atmosphere and to natural changes in global climate. The geologist can thus be called upon to be conversant in atmospheric chemistry and oceanography.

Earth sciences are in one sense simply branches of physics, chemistry, and biology. Yet the field is distinguished by virtue of the complexity of the target of study: Earth. Our planet and its solar system environs are extraordinarily complex systems, and the full spectrum of factors that can influence these systems is seldom, if ever, known. There is also the element of time. Like astronomy, earth sciences have a historical aspect that can involve unfathomable time scales. These distinctions can demand the liberal use of inference as a substitute for verification, a state of affairs that makes most scientists uneasy. The challenge of the twentieth century has been to ensure that the reliance on inference is kept to a minimum. Out of necessity then, earth scientists have both instigated and made use of new developments in the physical and life sciences.

Arguably, the most significant advance in earth science of the twentieth century was the discovery of plate tectonics. This all-encompassing theory brought order to seemingly unrelated observations made throughout the first half of the century. Driven by a combination of the efforts of insightful and dedicated scientists and technological advances born of war, plate tectonics ranks as one of the great scientific discoveries of the twentieth century.

Understanding Earth's Materials

Due largely to the work of U.S. chemist J. Willard Gibbs (b. 1839) and Austrian chemist and physicist Ludwig Boltzmann (b. 1844), the end of the nineteenth century saw the emergence of chemical thermodynamics as a fundamental branch of physical chemistry. With the advent of thermodynamics came the ability to ascertain the conditions under which minerals, the materials comprising rocks, form. Rather than simply describing rocks, scientists now had the theoretical framework with which to assess the physical and chemical significance of the rocks.

According to the theory of thermodynamics, combinations of minerals exist over a certain range of temperatures and pressures. Pressure increases with depth in the earth because of the weight of the overlying rock. This rise in pressure together with decay of radioactive elements causes temperature to rise with depth also. As pressure and temperature acting upon a rock are changed slowly by the rock's burial beneath growing mountains, the rock's minerals react with one another to form new minerals. Early in the twentieth century it was realized that the distribution of minerals exposed at the surface could be used to reconstruct the size and shapes of ancient mountain belts if the pressures and temperatures that caused them to grow were known. Rocks composed of minerals that grew at great pressures of thousands of atmospheres must have been at one time buried deep in the earth and then exhumed by erosion of the mountains above. In the years before World War II, Norwegian chemist Victor Moritz Goldschmidt (b. 1888) and Finnish chemist and geologist Pentti Eelis Eskola (b. 1883) showed that thermodynamics could be used to relate minerals at Earth's surface to the physical conditions of their growth. Through the efforts of the likes of Goldschmidt and Eskola, it became possible to decipher the evolution of ancient mountains belts.

Earth scientists not only made use of thermodynamics in the twentieth century, they improved upon it as well. Using x rays, mineral chemists could see that replacement of one element by another in a mineral did not alter its structure. The fact that their chemical compositions could be changed without affecting their structure meant that minerals had to be treated differently from the solutions with which most chemists were accustomed to working. Geoscientists adapted thermodynamics so that it could be used with these peculiar constraints imposed by crystalline substances.

Similarly, Russian chemist D. S. Korzhinskii (b. 1899) showed that traditional thermodynamics could not be used to predict which minerals will form in a rock when water flows through its pores. Because water commonly flows through rocks, even at considerable depths, in 1959 Korzhinskii devised ways of expanding the rules of thermodynamics to allow for reaction between rock and flowing fluids.

The Industrial Revolution afforded the technology to verify the concepts of thermodynamics in the laboratory. In the earth sciences this process began in earnest in 1905 when the Carnegie Institution of Washington, D.C., established the Geophysical Laboratory staffed with physicists, chemists, and geologists. It was here that Canadian geologist Norman Levi Bowen (b. 1887) and his colleagues revolutionized geology by establishing the chemical and physical state of molten and solid rock at high temperatures in the laboratory. Bowen's experiments allowed him to suggest that Earth's crust was the product of melting of parts of the mantle, a process known as differentiation. His book entitled *The Evolution of the Igneous Rocks*, published in 1928, is considered the most influential work on the origin of igneous rocks of the twentieth century and firmly established the potential of the physical chemical approach to geology.

The properties of minerals are the result of the different ways in which their constituent atoms are bonded together chemically. Understanding the behavior of rocks at high pressures and temperatures in the earth requires an understanding of materials at the microscopic level. Early in the twentieth century, U.S. chemist, biologist, and two-time Nobel laureate Linus Pauling (b. 1901) formulated the electronegativity scale and so provided a logical means for describing how atoms bond with one another. Pauling's first paper, in 1923, dealt with the crystal chemistry of the mineral molybdenite. The first diagram in his classical monograph *The*

Nature of the Chemical Bond (b. 1939) shows the crystalline structure of rock salt.

In *The Nature of the Chemical Bond* Pauling enumerated a set of rules for rationalizing the structures of crystalline substances (such as minerals). Pauling's rules, as they have come to be known, are still used as a first-order means for understanding mineral structures and his work ranks as one of the most important advances in understanding rock materials.

It was Pauling who elucidated the nature of the bond between silicon and oxygen, the fundamental building blocks of Earth and the other rocky planets. The bond between atoms of silicon (Si) and oxygen (O) is literally the glue that holds Earth together. In 1980 when his description of the Si-O bond was questioned on the basis of a more modern approach, erstwhile mineral chemist Pauling took the time to publish an article in the *American Mineralogist* to restate his case.

The early work on bonding laid the foundations for the modern discipline of computational mineral physics. Today mineral physicists use supercomputers to simulate forces on individual atoms and bonding at the scale of electrons. With such calculations they can predict the behavior of minerals at extreme conditions inaccessible to humans, including at the deep interior of the earth.

Crystals of sodium chloride (salt) under polarized light and magnified 200 times. (Corbis/Science Pictures)

An understanding of bonding is also useful for studying the chemical processes that occur at the interfaces between Earth's regolith, atmosphere, and hydrosphere. As a result, new subdisciplines like environmental surface chemistry have evolved. Mathematical models for atomic and molecular bonding are used to predict the rates at which atoms in the environment adsorb (stick) and desorb (unstick) to mineral surfaces. Studies of this kind can be applied to better understand the rates of stream acidification and groundwater contamination and to devise methods for neutralizing contaminants. Development of new surface-sensitive analytical techniques like atomic force microscopy (AFM) are aiding scientists in the study of water-mineral interactions.

Earth's Deep Interior

At the close of the twentieth century, drilling deep into the earth is still the realm of science fiction. Engineering and financial constraints have limited the deepest holes to about 10 km/6.2 mi. As a consequence, direct examination of our planet is confined to the outer shell, which comprises less than 1 percent of its volume. Geophysics provides us with a "window" into the deep interior of Earth and that of other bodies of the solar system. Studies of Earth's magnetism, gravity, and seismicity—the traditional subdisciplines of geophysics— were all blossoming in the early part of this century.

Devices for measuring Earth's magnetic field were being towed across land and oceans around the world by 1905. As early as 1906 two distinct forms of rock magnetism had been discovered. Some rocks were magnetized with their north "poles" parallel to Earth's present magnetic field, while others were magnetized with their poles reversed with respect to Earth. In the 1960s this observation would prove to be crucial in the revolutionary discoveries of seafloor spreading and plate tectonics.

Geophysics as a tool for petroleum exploration began in 1922. By that time the association between salt domes and oil in the Gulf of Mexico was well known. Because salt is much less dense than rock, the gravitational field above salt domes should be distorted. The torsion balance, a device for measuring gravitational field strength, was used to locate salt domes, and thus oil, in the Gulf. For the first time, geophysics had been used for prospecting.

At about this same time man-made seismic waves were being tested as means for exploring the earth's shallow structures with an eye toward finding more oil, and by 1930 seismic reflection was well established as the most widely used geophysical tool for petroleum exploration.

Seismic waves resulting from earthquakes were being used to deduce Earth's layered structure in the earliest part of the century. Seismic waves, it was found, did not travel freely through the planet. Instead, Irish seismologist Richard Oldham (b. 1858) showed in 1906 that something blocked the waves from passing from one side of the globe to the other, resulting in a sort of seismic wave shadow. Oldham showed that the culprit was a core that behaved like a fluid. In 1914 German-born U.S. seismologist Beno Gutenberg (b. 1889) calculated that the surface of the newly discovered core lay at a depth of approximately 2,900 km/1,802 mi, and in 1936 Danish seismologist Inge Lehmann suggested that there was a solid core within the liquid core. Although circumstantial evidence for the presence of a solid inner core abounded, definitive seismological evidence for its existence did not come until 1998.

By the middle of this century the velocities of seismic waves revealed even greater detail about the layered structure of our planet. The speed with which a seismic wave moves through rock depends upon the density of the rock. The greater the density, the greater the velocity. It was known that there were several planetwide changes in seismic velocity with depth. The shallowest, the Mohorovicic discontinuity, named in honor of Croatian seismologist Andrija Mohorovicic (b. 1857), lies 5–10 km/3.1–6.2 mi beneath the ocean crust and approximately 35 km/22 mi beneath the crust of the continents. Below the "Moho" seismic waves travel nearly 20 percent faster than above and it is regarded as the bottom of Earth's crust.

In the last two decades of the century, three-dimensional imaging of Earth's seismic structure, a technique known as *global seismic tomography*, has provided snapshots of the mantle in much the same way that CAT scans are used to image the human brain in medicine. Using seismic tomography, geophysicists like A. M. Dziewonski of Harvard University and J. H. Woodhouse of Oxford University have shown that Earth's tectonic plates descend deep into the mantle, and that large portions of cold dense mantle slowly sink and are replaced by more buoyant warmer mantle. The ability to relate deep mantle structures to surface features of the planet constitutes a major advance in the earth sciences.

The earth is often described as being composed of an outer thin crust, a mantle, and a core. But as early as the 1920s it was known that there were other dramatic changes in seismic velocities that could not be explained by this simple picture. Velocities change dramatically with depth in the mantle at depths of 100–200 km/62–124 mi, and again at 400 and 600 km (248 and 372 mi). These transitions are every bit as fundamental as the distinction between the core, mantle, and crust, but their nature is the subject of continued research. The change in seismic velocity at about 600 km/372 mi depth is used to define the transition between the upper and lower mantle.

In order to relate seismic velocities, the window into Earth's deep interior, to the unseen materials that make up our planet at depth, earth scientists of this century have turned to the emerging field of high-pressure physics. Deep in the earth rocks are subjected to crushing pressures due to the burden of the weight of overlying rock. At the bottom of the crust, pressures are 20,000 times that exerted by our atmosphere. Physical chemistry suggests that the mineralogy of a rock will change in response to these immense pressures. High-pressure physics has provided the means for assessing what those changes are likely to be.

At the turn of the century the maximum pressure obtainable in the laboratory was on the order of 2,000 atmospheres. In 1910 U.S. physicist Percy Bridgman (b. 1882) invented a device, called the collar, that allowed him to confine materials between two pistons. With his new device Bridgman was able to squeeze all kinds of materials to pressures comparable to the base of Earth's crust. He discovered, for example, that he could squeeze water to produce a form of ice that could exist at temperatures near the boiling point of ordinary liquid water. This "hot ice" produced a sensation in the media.

Following World War II the pursuit of the first man-made diamonds led to rapid developments in high-pressure technology. Inexpensive diamonds would be invaluable for machining weaponry. While exploring ways of producing high temperatures and pressures, U.S. chemist Loring Coes, a researcher at the Norton Company in the United States, learned how to grow minerals at very high temperatures and pressures that simulated Earth's mantle. Coes could grow beautiful minerals like red garnets—minerals thought to be produced deep in the earth at high pressures. Geoscientists took note, and soon they too learned how to mimic Earth's pressures and temperatures in the laboratory. Coes discovered a high-pressure form of ordinary beach sand that now bears his name, coesite. Natural coesite was later discovered to be present in Meteor Crater, Arizona, where it had been produced by the great pressures caused by a meteor impact. Coes's research showed that new abrasives of economic value could be produced at high pressures. In 1954 researchers at General Electric in the United States succeeded in producing the first synthetic diamonds.

The U.S. military fostered high-pressure research in other ways too. Investigations into how metals behave at high pressures were useful for designing submarines. The Chicago University Institute of Metals

A giant meteor crater in a desert in Arizona. (Corbis/Charles and Josette Lenars)

designed a new pressure device made of a single diamond with a hole drilled in it. Diamond was used because it is made of carbon. Carbon is relatively transparent to x rays, and these researchers wanted to pass x rays through metals while they were being squeezed to examine how their structures changed. This simple device evolved into the diamond-anvil cell, in which samples are squeezed between two diamonds using a small hand-sized anvil. The principle is a simple one; a small force applied to a tiny area like the head of a cut diamond produces very large pressures.

With the diamond-anvil cell, physicists, chemists, and earth scientists would be able to produce pressures exceeding a million atmospheres. Indeed, this milestone was achieved by two U.S. earth scientists, David Mao and Peter Bell of the Carnegie Institution of Washington's Geophysical Laboratory in 1975.

In 1974 Australian earth scientist John Liu discovered that most of the earth is likely composed of a mineral with a structure wildly different from the minerals found near the surface. The majority of rocks are made from silicon chemically bound to oxygen. With very few exceptions, the tiny silicon atoms are surrounded by four larger oxygen atoms in minerals that makeup familiar rocks like granite. But Liu's new mineral, magnesium silicate perovskite, is different because the small silicon atoms are surrounded by six oxygens rather than four. This means that magnesium silicate perovskite is very much denser than the minerals familiar to geologists. Liu made his discovery by squeezing minerals thought to have the chemical composition of the mantle with the diamond-anvil cell while heating the mineral with a laser beam. Amazingly, the pressure at which the transition to perovskite occurred corresponded nearly exactly to the depth of the transition between the upper and lower mantle defined by seismic velocities. Liu's discovery showed that the increase in seismic wave velocities at the transition between the upper and lower mantle was caused by the conversion of "normal" minerals, in which four oxygens surround silicon, to minerals like perovskite, in which six oxygens are tightly packed around each atom of silicon. This work sparked a new era of cooperation between seismologists and high-pressure researchers in the earth sciences.

The success of the diamond-anvil cell has broadened the scope of high-pressure earth science. High-pressure researchers at institutions like the Geophysical Laboratory and the Bavarian Research Institute of Experimental Geochemistry and Geophysics (Bayerisches Geoinstitut) in Germany are going beyond conventional boundaries of the geological sciences to study the interiors of other planets in our solar system.

The giant planets like Jupiter and Saturn are composed primarily of hydrogen, the most abundant gas of interstellar space. What must these planets be like deep in their interiors, where pressures exceed a million atmospheres? In 1988 it was reported that a solid form of hydrogen had been obtained by compressing hydrogen gas to pressures of between one and two million atmospheres. This solid appeared to be metallic. Although the validity of the particular claim is debated, the implication is clear; the interiors of giant planets may be composed of metallic hydrogen.

The melting temperature of iron at very high pressures is another example of fundamental research into the behavior of materials with implications for planetary interiors. Surprisingly, we still do not know the temperature of Earth's core. Because the core is thought to be mostly made of iron, and the outer core is liquid, the melting temperature of iron would prove invaluable for estimating core temperature. At first, determining the temperature at which iron melts seems trivial. After all, molten iron has been used to produce steel since before the turn of the century. Surely, one might think, all that is necessary is an accurate measurement of a batch of molten iron in some foundry. But melting temperatures change with pressure, and so they must be measured at core pressures in order to be of use. This is no easy task and at present there are conflicting results obtained with the diamond-anvil cell.

Obtaining accurate estimates of the temperature of the core is a prerequisite for understanding the physics of the planet as a whole. Earth—unlike Mars, which ran out of heat long ago—is still geologically active because of the heat contained within. If we are to understand what causes the tectonic plates to move, for example, we must have a better understanding of the temperature of the core.

Geological Time

In the year 1900 the age of the earth was estimated by comparing the amount of salt in the oceans to the rate at which salt was being delivered to the seas by rivers. The calculation suggested that Earth was approximately 90 million years old. Discoveries of radioactivity and isotopes changed forever our view of geological time.

By 1917, due largely to the work of British physicist Ernest Rutherford (b. 1871), a coherent picture of the atom was emerging. The mass of the atom lay in its nucleus and its chemical behavior was a function of the tiny electrically charged electrons that surround the nucleus. The number of neutral particles in the nucleus, called neutrons, was found to be variable, giving rise to different masses (or weights) of the same element. The different masses of a given element were dubbed isotopes. Radioactivity, discovered just prior to the turn of the century, was rightly identified as the result of decay of an isotope of one element to form another element.

In 1905 U.S. chemist Bertram Boltwood (b. 1870), working with Rutherford, suggested that lead was the product of uranium decay, and in 1907 Boltwood set out to show that the relative numbers of parent uranium isotopes and the lead isotopes they produce could be used to determine the ages of uranium-bearing minerals. Using lead-to-uranium ratios, Boltwood showed that ages of rocks from several settings varied from 410 million to 2.2 billion years. These dates indicated an antiquity not previously considered. The science of isotope dating was born.

Numerous isotope decay series have been used to date rocks since the time of Boltwood, including the decay of rubidium isotopes to produce strontium, and the decay of potassium to produce argon. Over the past two decades, new methods like heating tiny mineral grains with lasers to release argon have drastically reduced the amount of material required for dating. The solid Earth is now known to be approximately 4,500 million years old, and uranium-lead dating of meteorites has allowed geoscientists to conclude that the solar system began to form from an interstellar cloud of dust and gas some 4,566 million years before present.

The Chemistry of Earth's Near Surface

The term *isotope* invariably conjures up the specter of radioactivity to nonscientists, but most chemical elements are composed of several isotopes. The majority of these isotopes, especially for the lighter elements, are not radioactive. The relative abundance of the so-called stable isotopes has proved to be invaluable in the modern earth sciences as a means for following the movements of the chemical elements back and forth between Earth's atmosphere, hydrosphere, biosphere, rocky crust, and regolith.

In 1931 U.S. chemist Harold C. Urey (b. 1893) discovered deuterium, a heavy isotope of hydrogen, ushering in the modern field of stable isotope geochemistry. Urey and colleagues later succeeded in isolating the rare heavy isotopes of several elements, including oxygen, carbon, nitrogen, and sulfur. Although chemically similar, isotopes of these lighter elements were found to be separable on the basis of mass alone. For example, water molecules composed of heavy isotopes of oxygen are less likely to boil away from a body of water by evaporation than are water molecules composed of the more abundant light isotope of oxygen. Thus rainwater, having been derived from evaporation of the oceans, has on average more of the light isotope of oxygen.

Because they can be separated by processes like evaporation, condensation, and chemical reaction, the stable isotopes of an element are like labels that can be used to trace elements as they pass back and forth from one part of the earth to another. Moreover, quantum mechanics shows that the amount of the rare heavy isotopes of elements like carbon, oxygen, and nitrogen relative to the amount of their more abundant light isotopes changes in predictable ways with temperature. On this basis Urey wrote a seminal paper in 1948 in which he foretold the use of stable isotopes as "geothermometers" and chemical tracers in the earth sciences over the next 50 years.

The abundance of isotopes of oxygen and carbon in the hard parts of marine organisms past and present are used routinely in the earth sciences to deduce temperatures in the oceans, and these same isotopes are being used to reconstruct the diets, physiology, and ecology of long extinct creatures. Measurements of stable isotopes in teeth and bones give clues as to whether mammoths were hunted to extinction by humans or died out because of changes in global climate. Isotopes may even tell us if dinosaurs were warm-blooded or cold-blooded. Indeed, an entirely new field, known as biogeochemistry, has evolved around the ability to use stable isotopes as a means for monitoring how chemical elements pass back and forth between living organisms and Earth's hydrosphere, atmosphere, and lithosphere.

Since the time of Urey it had been thought that the separation of two isotopes was always in proportion to the difference in molecular weight. In the early 1980s U.S. chemist Mark Thiemens and his colleagues performed experiments in the laboratory that proved otherwise. They demonstrated that the production of ozone by the joining together of three atoms of oxygen in the upper atmosphere causes separation of the two different heavy isotopes of oxygen relative to the one lightest isotope *independent* of their masses.

Thiemens immediately suggested that other reactions might behave similarly, and that these so-called symmetry induced isotope effects could explain the strange abundance of oxygen isotopes seen in some very old objects within stony meteorites. All at once Thiemens's discovery grabbed the attention of geochemists, cosmochemists, and atmospheric chemists. Today symmetry induced isotope effects are being used to track mixing in Earth's atmosphere. Analysis of Martian meteorites suggests that similar reactions may take place on Mars. Debate surrounding the significance of anomalous abundances of oxygen isotopes in meteorites continues.

Technological advances like refinement of accelerator mass spectrometer methods and laser sampling allow earth scientists to make use of

General Electric researchers use a mass spectrometer to isolate uranium (ca. 1940). (Corbis/Schenectady Museum; Electrical History Foundation)

more and more isotopes. Cosmogenic nuclides are isotopes produced as a consequence of bombardment by cosmic rays. Cosmogenic nuclides that decay rapidly, like the radioactive form of beryllium, are being used with increasing frequency to date recent fault movements and to trace past ocean circulation patterns.

Climate and Global Warming

At the start of the twentieth century, Swedish chemist and Nobel laureate Svante Arrhenius (b. 1859) suggested that changes in the amount of carbon dioxide gas in the air have influenced Earth's climate over time. The chemical bonds between carbon and oxygen in carbon dioxide cause these gas molecules to vibrate at low frequencies corresponding to infrared radiation, or radiant heat. The result is that heat from Earth's surfaces, warmed by high-frequency ultraviolet light from the sun, is absorbed by the vibrations within carbon dioxide molecules in the air and so cannot escape. The trapping of heat by atmospheric gases is referred to as the greenhouse effect.

Arrhenius's greenhouse effect has become a subject of intense study in the earth sciences in recent decades because of an increase in average global temperature of approximately 1°F (0.5°C) over the past century that has come to be known as global warming. Recent melting and collapse of the Larsen Ice Shelf of Antarctica is one obvious consequence of global warming. By studying the geological and historical records, earth scientists have established that global temperature has been highly variable in Earth history and that fluctuations in global temperature have occurred even in historical times. The recent rise in temperature is therefore not in itself unusual. However, it happens to coincide with the spread of industrialization, giving rise to the speculation that it has resulted from a man-made greenhouse effect in which Earth's radiant heat is trapped by atmospheric pollutants, especially carbon dioxide gas. At issue is whether or not global warming is anthropogenic—resulting from man-made gases in the atmosphere—or simply a natural fluctuation like so many others in Earth history.

Earth scientists must consider, for example, that the present episode of global warming has thus far still left England approximately 1°C cooler than during the peak of the so-called Medieval Warm Period (A.D. 1000–1400). The latter was part of a purely natural climatic fluctuation on a global scale. With respect to historical times, the interval between the Medieval Warm Period and the rise in temperatures we see today was unusually cold throughout the world. Some scientists argue on this basis that, rather than being an anomaly, the present warming is a return to a more "normal" state. Assessing the impact of humankind on global climate is complicated by the natural variability on both geological and human time scales.

In order to evaluate the likely influences of recent changes in the amounts of greenhouse gases in our atmosphere, earth scientists require a record of fluctuations in atmospheric carbon dioxide in the past. Tiny samples of air trapped in polar ice provide one such record for the more recent historical past. The ice preserves annual layering that can be counted like tree rings to obtain the age of each layer. Trapped bubbles of air thus provide a record of how the chemistry of the atmosphere has evolved over the lifetime of the ice.

At the close of the twentieth century, improved understanding of the geochemistry of carbon is providing the means for correlating carbon dioxide in air to climate change over geological time scales. Assessing the role of carbon dioxide as a cause for past fluctuations in climate requires a thorough understanding of how carbon moves in and out of the atmosphere. It is still unknown, for example, to what extent the oceans can mitigate increases in atmospheric carbon dioxide. Because the processes involved are slow in human terms, one of the most effective ways of understanding the slow exchange of carbon between Earth's atmosphere, hydrosphere, biosphere, and lithosphere is through examination of the chemistry of carbon as recorded in the geology of the planet. Results of such studies in the last decade have shown that the amount of carbon dioxide in the atmosphere today is unusually low in comparison to most of the rest of the last 600 million years. The onset of the ice ages, times of advance of ice sheets across the continents that began approximately one million years ago, has been attributed to the small amount of carbon dioxide in the atmosphere and the average cooling that should result. The link between atmospheric carbon dioxide concentration and ice ages is bolstered by these same studies showing that the only other time of significant continental glaciation, about 300 million years ago, was also a time when the amount of carbon dioxide in the atmosphere was anomalously low, like today.

The connection between short-lived changes in climate, on the order of a single year, and catastrophic geological phenomena like volcanic eruptions was known as far back as the eighteenth century. In the last two decades it has become clear that longer-term changes in global climate are influenced by fundamental processes like mountain building and mineral reactions at great depths in the earth. Today, for example, it is suggested that the single most influential factor in global climate over the past 50 million years has been the rise of the Tibetan Plateau.

Long-term changes in climate can be related to changes in Earth's orbit relative to the sun. In the early part of the century it was known that wobble of Earth's rotation axis occurs with a periodicity of 22,000 years, that variations in axis tilt repeat every 40,000 years, and that ellipticity, or eccentricity, of the orbit changes with a periodicity of about 100,000 years. In 1920 Yugoslavian meteorologist and mathematician Milutin Milankovitch (b. 1879) showed that the amount of energy, or heat, received by Earth from the sun varies with these long-term changes in orbit. Changes in solar radiation with orbit are now known as Milankovitch cycles, and their capacity for altering Earth's climate has been the subject of study ever since their discovery. Over the next 20 years Milankovitch and others promulgated the notion that the advance of ice sheets during the ice ages and their subsequent retreats might be correlated with Milankovitch cycles. Special emphasis was placed on correspondence between 100,000-year cyclicity in the ice ages and orbital eccentricity. Decades later, using the isotopes of oxygen in remains of marine organisms as "paleothermometers," scientists correlated changes in ocean temperatures over hundreds of thousands of years to Milankovitch cycles. Periodic changes in Earth's orbit due to interactions among bodies of the solar system is still not fully understood, and the influence of orbital forcing on cyclical climate change remains an important area of research in the earth sciences at the close of the twentieth century. Most recently it has been suggested that periodic changes in the amount of interplanetary debris, principally dust, encountered by Earth as it travels around the sun can affect our climate.

Life on Earth

Studies of the origin and evolution of life are, perhaps above all other scientific pursuits, truly interdisciplinary, requiring synergetic relationships among geologists, biologists, biochemists, and organic and inorganic chemists.

Metazoa, or animals, first appeared in the fossil record approximately 680 million years before present. The abrupt appearance of multicellular animals implied that early life had evolved at an unrealistically rapid pace. Scientists of the late nineteenth and early twentieth centuries were thus prompted to search for evidence of more primitive forms of life in older, apparently barren rocks. The hunt produced some scant fossil evidence for very ancient life, but it was not until the discovery in 1954 of a plethora of diverse fossil microscopic organisms in the Gunflint rocks along the north shore of Lake Superior in the United States that the existence of premetazoan life became widely accepted. The Gunflint biota proved that there had been significant evolutionary activity by at least around 2,000 million years before present. In the decades that have followed, studies of stromatolites, which are rocks made from ancient sediments laid down on sticky mats composed of microbes, have shown that life existed at least as far back as 3,600 million years ago.

In the late 1990s measurements of carbon isotope abundances in graphite from a rock formed nearly 3,800 million years ago are consistent with the existence of life. The age of the rock is, however, in question, and evaluation of this evidence for ancient life requires reference to the geology of the solar system as well as the biogeochemistry of Earth. Studies of craters that abound on the moon and on the rocky surfaces of other bodies of the solar system show that 3,800 million years ago impacts by asteroid-sized objects—vestiges of planet formation—were commonplace. Life, it is argued, was unlikely to have survived under such hostile conditions.

Plants produce oxygen through photosynthesis. The geological record shows that oxygen was not always abundant in our atmosphere. When did Earth's atmosphere become enriched in oxygen? Was it the result of photosynthesis or the exhaustion of the supply of iron that was being rusted by reaction with oxygen? Answers to these questions require a synthesis of paleobiological and geological data. For example, in order to evaluate the influence of life on Earth's atmosphere, it is necessary to have some understanding of the atmosphere before there was life. One possibility is that the atmosphere was exhaled from volcanoes. Geochemists thus study rocks from which the primordial atmosphere could have been expelled so that they can determine the amount of oxygen that might have existed prior to photosynthesis.

Meteoriticists ponder the effects that bombardment by asteroids and comets might have had on Earth's early atmosphere. It has been suggested, for example, that the water necessary to form oceans and life itself may have come from the melted ice of comets that struck the earth more than 4 billion years ago.

The origin of organic compounds on our planet has yet to be resolved, despite intensive and creative study over the past several decades. In 1953 then graduate student Stanley Miller, working with U.S. chemist Harold Urey, performed an experiment that showed how organic compounds might have begun on Earth. They exposed a warm, gaseous mixture of inorganic compounds, including ammonia, methane, hydrogen, and water, to electrical discharges. The gases were meant to represent the prebiotic atmosphere, the proverbial primordial soup, and the electrical discharges simulated lightning. After several days Miller and Urey produced amino acids, the ubiquitous building blocks of life that allow chemical reactions to take place in the cells of living things. Yet the Miller-Urey reactions are not the only source of amino acids. They are also abundant in certain kinds of primitive stony meteorites and similar objects that undoubtedly struck the earth with comparative frequency early in its history. Could life have begun as a result of the bombardment by organic-rich asteroids?

Toward the end of this century, the capacity to separate organic compounds by various forms of chromatography permitted studies of fossil life to be extended to the molecular level. As more amino acids were identified in ancient materials, the method of racemization age dating was developed. Amino acids can exist in two forms, one being a mirror image of the other. In living organisms only one form exists. Upon death, the living form transforms spontaneously to its mirror image at a steady rate over time. The amount of the living form relative to the racemized form thus provides an estimate of the time since death. Organic-bearing fossils can be dated as far back as 20 million years with this method.

The geochemistry of organic compounds has provided new insights into the fate of biota after death. U.S. biogeochemist T. C. Hoering showed that carbon isotopes of organic compounds extracted from ancient rocks could be used to learn about the processes that give rise to petroleum.

The phrase "you are what you eat" applies to the isotopes of both modern and ancient organisms. Increasingly, biogeochemists are able to reconstruct the food chains of creatures ranging in complexity from bacteria to elephants and even humans. Comparisons of the isotope effects of metabolic processes in modern and ancient biota is an evolving tool in the earth sciences. For example, it has been shown that the relative abundances of the stable isotopes of carbon from modern blue-green algae that thrive in hot springs rich in carbon dioxide are similar to those preserved in ancient, Precambrian stromatolites. The implication is that Earth's atmosphere was richer in carbon dioxide in the Precambrian than it is today. Such conclusions would not have been possible prior to the confluence of technology and expertise from a variety of scientific specialties that is, for the lack of a better term, a subdiscipline of biogeochemistry.

Considerable attention was drawn to the phenomenon of mass extinction in the 1980s and 1990s. This attention arose from the bold assertion made in 1980 by Nobel Prize–winning physicist Luis W. Alvarez (b. 1911) and his colleagues that the dinosaurs and 70 percent of all other species of the earth were wiped out 65 million years ago by the impact of an asteroid or comet measuring approximately 10 km/6.2 mi or more in diameter. Alvarez and his coworkers made their proposal on the basis of an unusual enrichment in the rare element iridium, an element chemically similar to platinum and gold, in a layer of clay marking the boundary between the Cretaceous and Tertiary geological periods. The latter is referred to as the K-T boundary. Since their original report, the iridium anomaly, as it has come to be known, has been found in rocks and sediments deposited at the K-T boundary around the world. In 1991 a circular impact structure of K-T age was found buried beneath 1 km/0.62 mi of carbonate rock in Mexico's Yucatan Peninsula. The structure, the Chicxulub crater, is the best candidate for the K-T impact site envisioned by Alvarez.

Despite its obvious attraction to scientists and nonscientists alike, the Alvarez impact hypothesis is not without problems, and there is another catastrophe that might well explain the K-T mass extinction. Massive volcanic eruptions in what is now the Deccan plateau region of western India coincided with the K-T extinction. The environmental effects of large-scale volcanism are every bit as menacing to life as those of an asteroid collision.

New theoretical studies suggest that mass extinctions are to be expected as a consequence of the dependence of living organisms on one another. Mathematical models constructed on this basis successfully mimic the pattern of extinctions over geological time, and suggest that mass extinctions are caused when vulnerability to extinction by normal evolutionary change is accompanied by unusual environmental stress imposed by catastrophes. Most scientists agree that a large asteroid-like body struck the earth near the present-day Yucatan Peninsula during K-T boundary. There is less agreement as to whether this event by itself killed the dinosaurs.

It is estimated that, on average, asteroids or comets measuring 5–10 km/3.1–6.2 mi in diameter strike the earth every 50 to 100 million years. Debate surrounding the Alvarez hypothesis has heightened awareness of the influence of asteroid or comet impact on Earth's geological history.

Space exploration, biology, and geology have joined forces several times in this century. The return of samples from the moon beginning in 1968 resulted in an infusion of funding for the development of equipment capable of detecting living organisms that might exist at extreme conditions. With the suggestion in 1997 that meteorites that most probably come from Mars may show evidence for microbial life, the scientific community is again tooling up to intensify the search for life at extreme conditions.

What is meant by life at extreme conditions? In 1977 scientists from the project FAMOUS (French-American Mid-Ocean Undersea Study) set out to observe firsthand the process of seafloor spreading. Upon descending in their deep-sea submersible vehicle, named *ALVIN,* to depths of 2–3 km/1.2–1.8 mi, the scientists were startled to find a host of new and

Deep-sea tube worms. (Corbis/Ralph White)

strange life forms. These creatures were concentrated near undersea hot springs heated by the volcanism that drives seafloor spreading. Perhaps the strangest of these creatures were the large red and white tube worms. Subsequent studies showed that these life forms owe their survival to symbiosis with bacteria that thrive on the hydrogen sulfide gas (familiar to many of us by its rotten egg odor) emanating from the volcanic hot springs. Here, the energy for life comes from chemical processing of volcanic gases rather than from the sun's ultraviolet rays. Discovery of these organisms showed that life could exist at higher temperatures, higher pressures, and with less light than previously thought—that is, at extreme conditions.

If life existed on Mars, and this is far from proved, then could it still exist today beneath the surface of the planet? Answers to these questions begin at home. It is not yet known with certainty how deep into Earth's regolith life exists, nor is the maximum temperature at which microbial life can exist on Earth. Terrestrial geologists, planetary geologists, microbiologists, and biochemists are actively pursuing answers to these questions.

Plate Tectonics

Many of us at one time or another have glanced at a globe or a map of the world and noticed that the eastern edge of South America fits neatly with the western edge of Africa. So too have many serious thinkers before us. And many of us have been exposed to the idea that there once was a great continent, and that the parallelism of the continental coastlines is no accident. In hindsight this seems obvious. But it is useful to explore this form of logic further in order that we might better understand the boldness required to suggest that the continents of Africa and South America, for example, were indeed once attached.

Upon further examination of maps of our world one notices that more than a few of the major land masses appear to "drip" southward, as if someone had poured a can of brown paint over the top of the blue planet. One can observe, for example, that the narrow ends of Africa, South America, North America, India, and lands bordering the Mediterranean Sea all point southward. According to the well-established tenets of physics, gravity pulls these masses of rock toward the center of the earth. The shapes of the land masses must therefore be attributed to a misleading coincidence, something not uncommon in nature. Who among us would be willing to suggest a force capable of stretching the land masses of the planet southward in the face of the physical laws that require that no such force exists? Such was the dilemma of scientists at the turn of the century as they tried to explain away the resemblance of the continents to pieces of a tightly fitting jigsaw puzzle.

Although several scientists of the nineteenth and early twentieth centuries had suggested that the continents were once joined and then drifted apart, German meteorologist Alfred Wegener is credited with having put forward a coherent theory for continental drift backed by the strength of his conviction. He suggested that approximately 200 million years ago a single great continent, which he called Pangaea, began to break apart. Wegener was influenced not only by the jigsaw fit of coastlines, but also by paleontological evidence suggesting that there must have been a bridge of land between Brazil and Africa, and between Australia, Africa, and India. In addition he cited abundant similarities in the geological histories recorded by rocks on either side of the Atlantic. He put forth his ideas on "Die Verschiebung der Kontinente" ["Continental Displacement"], first in a lecture in 1912 and then in a series of publications from 1913 to 1924.

Perhaps the greatest of all insults to a scientist is to be ignored. If so, Wegener's hypothesis landed him many a compliment. His paleontological and geological evidence was scrutinized and seemingly refuted. A few prominent geologists could see merit in Wegener's *continental drift,* as it came to be known in English. But the fatal flaw in his hypothesis was the lack of an explanation as to how continents could plow through the dense rock of the ocean basins without complete demolition.

We now know that the solution to Wegener's problem was that continents do not push through the oceans like icebreakers, but instead ride passively atop plates that include the ocean basins. In a remarkably prescient work in 1929, British geologist Arthur Holmes (b. 1890) described a mechanism for continental drift that bears many similarities to present-day plate tectonics.

Holmes had studied radioactivity in rocks for years, and it was well known that radioactive decay causes rocks to heat up. He used his knowledge of the amount of heat in the earth to surmise that at great depth rocks must "flow" much like the way hot water rises, cools, and then sinks in a boiling pot, a process known as convection. Holmes argued that the earth was layered. The upper brittle portion composed of rocks like granite was underlain by a "fluid" layer that, given sufficient time, would flow. A useful analogy can be made between flow of Holmes's "fluid" layer over millions of years and the way in which old windowpanes flow, and thus thicken at their base, over hundreds of years. This so-called fluid layer was composed, it was asserted, of peridotite, a rock denser than granite.

Because continents were known to be rich in radioactive elements like uranium, Holmes concluded that temperature was highest beneath continents. The hot rock beneath was apt to rise during convection and thus force the continents apart. Conversely, at the margins of continents the underlying rock would tend to sink , explaining the presence of deep oceanic trenches around the rim of the Pacific, for example. With his model Holmes was able to offer cogent explanations for many geological phenomena, including mountain building and volcanism in Japan, the Aleutian Islands, and around the rest of the Pacific margin.

Lack of new information slowed the pace of ideas concerning continental drift until after World War II. As a result of the war, sonar became available to better map the topography of the ocean bottom. Devices called magnetometers, perfected for use as submarine chasers, could be used to detect more accurately variations in Earth's magnetic field. By the mid-1950s these technological advances afforded new data that begged explanation.

Mapping of the ocean floor showed that there was a semicontinuous mountain range, or mid-ocean ridge, 50,000 km/31,068 mi long beneath the seas. The average elevation of this mountain chain above the seafloor is, it was discovered, nearly as high as the tallest mountain in North America. The ridges spew large amounts of heat and are rocked by earthquakes.

Deep canyons at the bottom of the sea were discovered too. These deep-sea trenches were found concentrated along the margins of the Pacific Ocean basin seaward of the many volcanic centers that comprise the so-called ring of fire, including the Philippine and Aleutian Islands and Japan. The trenches are also seismically active.

Surveys of rock magnetism showed that either Earth's magnetic field or the rocks that record it were moving across the globe with time. Meanwhile, the evidence for periodic reversals in Earth's magnetic field was mounting; by accurately dating magnetized rocks using the amount of radioactive decay of potassium (K) to form argon (Ar), it was shown that rocks of similar age from widely separated regions showed the same magnetic polarity.

Princeton University geologist Harry H. Hess (b. 1906) put forth a powerful explanation for the new postwar observations in a widely circulated but unpublished manuscript in 1959. During World War II Hess captained a U.S. Navy ship in the Pacific. Between battles, he conducted echo soundings of the seafloor and his attention was drawn to explaining the structure of the ocean basins. His conclusion, elaborated in a landmark paper in 1962 entitled "History of Ocean Basins," was that the ocean crust, made of basalt, a dark-colored volcanic rock, was like a giant conveyor belt. It was produced by volcanism at the mid-ocean ridges, pushed or pulled away from the ridge axis, and eventually destroyed by plunging down into the deep-sea trenches. Hess's hypothesis, influenced by the earlier concepts of Arthur Holmes and referred to as seafloor spreading, was embraced by the scientific community.

In 1965 Canadian geologist J. Tuzo Wilson elaborated on Hess's seafloor spreading idea. He suggested that the mid-ocean ridges, deep-sea trenches, and the faults that connect them combine to divide the earth's outer layer into rigid independent plates. The connecting faults were shown by Wilson to have a unique geometry, and were called transform faults.

The formal theory of plate tectonics was enunciated in the late 1960s by Jason Morgan of Princeton University, Dan McKenzie of Cambridge, and Xavier Le Pichon, then at the Lamont Observatory in New York. All three had training in physics that benefitted the development of the theory. Morgan used mathematical arguments to show that the rigidity of the plates requires that they extend to depths beyond the lower limit of the oceanic crust. The term *tectosphere* was used to describe the crust and upper mantle that comprise the rigid plates. The tectosphere is now referred to as the lithosphere. Midoceanic ridges, or spreading centers, were identified as sites of plate construction. Trenches, or subduction zones, marked the places where the plates sink back into the deep earth. Le Pichon, McKenzie, and their colleagues demonstrated that motions of the plates made sense, geometrically, when inferred from the relative positions of the spreading centers, subduction zones, and connecting transform faults.

Geophysicists like McKenzie and Don Anderson of the California Institute of Technology used mathematics to investigate possible driving forces for the plates. Again, the ideas of Arthur Holmes proved prescient. Convection in the mantle, whereby hot rock creeps slowly upward, spreads latterly, and then cools and descends, seemed the best candidate. Upward convection was postulated to be occurring beneath the midocean ridge system, while downward convection explained the sinking of the plates at deep-sea trenches.

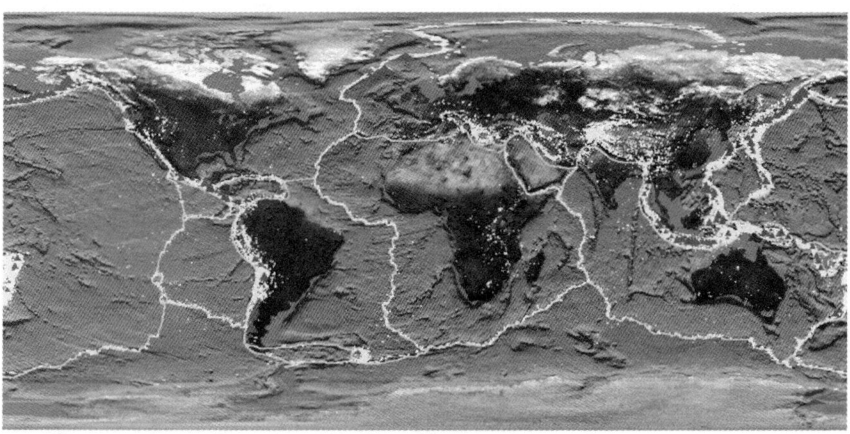

A map of the world with plates outlined in white. (NASA)

The successes of plate tectonics as a theory are vast. In 1963 Canadian L. W. Morley and, later, Fred Vine and Drummond Matthews of Cambridge University showed that reversal of Earth's magnetic field in combination with seafloor spreading could account for the "stripes" of magnetic polarity recorded in basaltic rocks of the ocean bottom. These stripes are symmetrically disposed about the mid-ocean ridges, and result from cooling of the newly made volcanic basalt as it is pushed or pulled away from the axis of the ridge. As the rock cools, it takes on the magnetic field of the earth at that time, and so forth. Patterns of volcanism and seismicity concentrated in mid-ocean ridges and deep-sea trenches around the globe are, to a first order, explained by seafloor spreading and plate tectonics.

We now know that the motions of the plates of the lithosphere are aided by the presence of a mobile layer of the mantle beneath them, known as the asthenosphere. The asthenosphere, defined by the slowing of seismic waves as they pass through it, is remarkably similar to Harry Hess's "fluid" layer.

The word *tectonic* connotes assembly, and plate tectonics derives its name in part because the distribution of mountain ranges is explained by the motions of the plates of the lithosphere. As the Indian plate moved northward, the continental crust riding passively on it eventually encountered the continental crust of the Eurasian plate. Both continental masses are too buoyant to be pulled down with the remainder of the subducting lithosphere and so they are squeezed together. Thus the Himalayan mountain range was born.

Beyond Plate Tectonics

Beginning in about 1980 many geologists, and particularly those working in western Canada, began to find evidence that some mountain ranges are composed of numerous small pieces of crust that originate from far-off places. A new terminology ensued. Blocks of crust that are bounded by faults on all sides were labeled suspect terranes by some scientists because their place of origin was often in doubt, or suspicious. If and when a fault-bound block is shown to be far traveled, its status is upgraded from a suspect terrane to simply terrane. The terrane concept predicts that Borneo, Sumatra, and Java may be future terranes to be imbedded into the margin of a larger continent following millions of years of plate motions.

Although it explains the positions of the mountain ranges, the ways in which continents deform to form mountains are not explained by plate tectonics. Fundamental observations like the heights of mountains on continents remain unexplained. Today, researchers are using advancements like the global positioning system (GPS) to measure the deformation of continents with great precision. The results of such studies may one day be useful for identifying areas on the continents that are susceptible to earthquakes.

—*Edward D. Young*

References

Bengston, Stefan, ed. 1994. *Early Life on Earth*. New York: Columbia University Press. 630 p. ISBN 0231080883.

Briggs, Derek E. G., Douglas H. Erwin, and Frederick J. Collier. 1994. *The Fossils of the Burgess Shale*. Washington, DC: Smithsonian Institution Press. 238 p. ISBN I560983647.

Hallam, A. 1973. *A Revolution in the Earth Sciences*. Oxford, UK: Clarendon Press. 127 p. ISBN 0198581459.

Hazen, Robert M. 1993. *The New Alchemists, Breaking through the Barriers of High Pressure*. New York: Times Books, Random House. 286 p. ISBN 0812922751.

Hoefs, Jochen. 1997. *Stable Isotope Geochemistry*. Berlin: Springer. 201 p. ISBN 3540611266.

Liu, L.-G. 1974. "Silicate Perovskite from Phase Transformations of Pyrope-Garent at High Pressure and Temperature." *Geophysical Research Letters* 1:277–280.

Mao, Ho-Kwong, and Peter M. Bell. 1976. "High Pressure Physics: The 1-Megabar Mark on the Ruby R1 Static Pressure Scale." *Science* 191: 851–862.

McKenzie, D. P., and P. L. Parker. 1967. "The North Pacific: An Example of Tectonics on a Sphere." *Nature* 216:1276–1280.

Miller, S. L., H. C. Urey, and J. Oro. 1976. "Origin of Organic Compounds on the Primitive Earth and in Meteorites." *Journal of Molecular Evolution* 9:59–72.

Officer, Charles, and Jake Page. 1993. *Tales of the Earth, Paroxysms and Perturbations of the Blue Planet*. Oxford, UK: Oxford University Press. 226 p. ISBN 0195077857.

Patterson, C. 1956. "Age of Meteorites and the Earth." *Geochimica et Cosmochimica Acta* 2:230–237.

Pauling, Linus. 1960. *The Nature of the Chemical Bond, and the Structure of Molecules and Crystals: An Introduction to Modern Structural Chemistry*. Ithaca, NY: Cornell University Press. 644 p. ISBN 0801403332.

Wilson, J. Tuzo. 1963. "Continental Drift." *Scientific American* (April): 2–16.

Yoder, Hatten S., Jr. (1993) "Timetable of Petrology." *Journal of Geological Education* 41:447–489.

2 Chronology

1900

German meteorologist Wladimir Peter Köppen develops a mathematical system for classifying climatic types, based on temperature and rainfall. It serves as the basis for subsequent classification systems.

1902

New Zealand–born British physicist Ernest Rutherford and British physicist Frederick Soddy discover thorium X and publish *The Cause and Nature of Radioactivity,* which outlines the theory that radioactivity involves the disintegration of atoms of one element into atoms of another, laying the foundation for radiometric dating of natural materials.

1905

U.S. chemist Bertram Boltwood suggests that lead is the final decay product of uranium. His work will eventually lead to the uranium–lead dating method.

The Carnegie Institution of Washington establishes the Geophysical Laboratory.

1906

Irish geologist Richard Oldham proves that the earth has a molten core by studying seismic waves.

Two distinct forms of rock magnetism are discovered: Some rocks are magnetized with their north "poles" parallel to Earth's present magnetic field and others are magnetized with their poles reversed with respect to Earth's magnetic field.

1907

French physicist Pierre-Ernest Weiss develops the domain theory of ferromagnetism, which suggests that in a ferromagnet, such as lodestone, there are regions, or domains, where the molecules are all magnetized in the same direction. His theory leads to a greater understanding of rock magnetism.

German scientists Richard Anschütz and Max Schuler perfect the gyrocompass, which always points to true north.

U.S. chemist Bertram Boltwood uses the ratio of lead and uranium in some rocks to determine their age. He estimates his samples to be 410 million to 2.2 billion years old.

1908

May 13. State and territorial governors from 44 states attend an environmental conservation conference, convened by President Theodore Roosevelt, at the White House in Washington, D.C.

June. President Theodore Roosevelt appoints a 57-member National Commission for the Conservation of Natural Resources. The commission will be directed by Gifford Pinchot, who first applied the term *conservation* to the environment.

1909

U.S. geologist and secretary of the Smithsonian Institution Charles D. Walcott discovers fossils of soft parts of organisms in the Cambrian Burgess Shale of the Canadian Rockies. The discovery provides unprecedented evidence pertaining to the rapid evolution of life that started in the Cambrian period.

Croatian physicist Andrija Mohorovicic discovers the Mohorovicic discontinuity in the earth's crust. Located about 30 km/18 mi below the surface, it forms the boundary between the crust and the mantle.

The crew of the brigantine *Carnegie* stands in front of the binnacle, a glass dome protecting the ship's compass (ca. 1920). (Corbis/The Mariner's Museum)

June 12. The research sailing vessel *Carnegie* is launched. Built out of nonmagnetic materials to minimize errors, the ship will conduct magnetic, electric, and gravity surveys of the globe.

1910

U.S. physicist Percy Bridgman invents a device, called the *collar,* that allows him to squeeze all kinds of materials to pressures comparable to the base of Earth's crust, giving rise to the new fields of high-pressure physics and mineral physics.

The Swedish archeologist Gerhard de Geer publishes *A Geochronology of the Last 12,000 Years,* setting out his influential system for dating rock strata.

1912

German meteorologist Alfred Wegener suggests the idea of continental drift and proposes the existence of a supercontinent (Pangaea) in the distant past.

German physicist Max von Laue demonstrates that crystals are composed of regular, repeated arrays of atoms by studying the patterns in which they diffract x rays. It is the beginning of x-ray crystallography.

1913

English geologist Arthur Holmes's book *The Age of the Earth* is published. In it Holmes espouses the potential of radiometric dating and suggests an age of 1.3 billion years for Earth.

English physicist J. J. Thomson discovers neon 22, an isotope of neon. It is the first isotope of a nonradioactive element to be discovered.

English physicists William and Lawrence Bragg develop x-ray crystallography by establishing that the orderly arrangement of atoms in crystals displays interference and diffraction patterns. They also demonstrate the wave nature of x rays.

French physicist Charles Fabry discovers the ozone layer in the upper atmosphere.

1914

German-born U.S. geologist Beno Gutenberg discovers the discontinuity that marks the boundary between the earth's lower mantle and outer core, about 2,800 km/1,750 mi below the surface.

1915

The English physicists William and Lawrence Bragg win the Nobel Prize for Physics for their work showing that the atomic structure of crystals can be analyzed from the diffraction patterns of x rays.

Finnish chemist and geologist Pentti Eelis Eskola develops the concept of metamorphic facies.

1916

The Dutch chemist Peter Debye demonstrates that the powdered form of a substance can be used instead of its crystal form for the x-ray study of its crystal structure.

1918

The English physicist Francis Aston builds the first mass spectrograph, building on earlier work by J. J. Thomson, which allows him to separate ions or isotopes of the same element. The mass spectrometer will become an important tool in stable isotope geochemistry.

1920

Yugoslavian meteorologist and mathematician Milutin Milankovitch shows that the amount of energy, or heat, received by Earth from the sun varies with long-term changes in Earth's orbit. Decades later, scientists will correlate fluctuations in global temperature to his "Milankovitch cycles."

The English physicist Frederick Soddy suggests that isotopes can be used to determine geological age.

1922

Geophysics is used for prospecting for the first time. The torsion balance, a device for measuring gravitational field strength, is used to locate salt domes, and thus oil, in the Gulf of Mexico.

1923

The first seismic prospecting takes place in the United States when geologists use seismometers to discover an oil field.

The German-born British physicist Frederick Lindemann investigates the size of meteors and the temperature of the upper atmosphere.

1925

The U.S. Navy develops a pulse modulation technique to measure the distance above the earth of the ionizing layer in the atmosphere.

1926

The Dutch physicist Willem Keesom solidifies helium.

The Scott Polar Research Institute is opened in Cambridge, England, to conduct Antarctic research.

1928

U.S. geochemist N. L. Bowen publishes *The Evolution of the Igneous Rocks*, in which he suggests that Earth's crust is the product of melting of parts of the mantle, a process known as differentiation. His work firmly establishes the potential of the physical-chemical approach to geology.

1929

Fire destroys the research sailing vessel *Carnegie* in Samoa as the result of a gasoline explosion after the ship had logged 342,681 miles of ocean during magnetic, electric, and gravity surveys of the globe.

British geologist Arthur Holmes describes a mechanism for continental drift that bears many similarities to present-day plate tectonics.

Norwegian chemist Victor Moritz Goldschmidt produces the first table of ionic radii useful for predicting crystal structures.

By studying the magnetism of rocks, the Japanese geologist Motonori Matsuyama shows that the earth's magnetic field periodically reverses direction.

1930

June 11. The first bathysphere, a spherical steel craft for undersea exploration, built by U.S. zoologist William Beebe and U.S. engineer Otis Baron, descends to 435 m/1,428 ft.

December. Hundreds of people fall ill and 60 die during a four-day fog in the industrialized Meuse Valley in Belgium. It is the first recorded air pollution disaster.

The Indian physicist Chandrasekhara Venkata Raman receives the Nobel Prize for Physics for his work on the scattering of light and for the Raman effect. Raman spectroscopy later becomes an important tool in mineral physics.

The Woods Hole Oceanographic Institution is established in Massachusetts.

Explosion seismology—the study of seismic waves caused by explosions—is a well-established technique used by the oil industry to explore for oil in the United States.

1931

U.S. chemist Harold C. Urey discovers deuterium, a heavy isotope of hydrogen, ushering in the modern field of stable isotope geochemistry.

1934

August 18. U.S. explorers and biologists William Beebe and Otis Baron descend in a bathysphere to a record 923 m/3,028 ft in the Atlantic off Bermuda.

William Beebe and Otis Barton pose with their invention, the bathysphere, in Bermuda (ca. 1934). (Corbis/Ralph White)

1935

U.S. seismologist Charles Richter introduces the Richter scale for measuring the magnitude of earthquakes at their epicenter.

1936

From the study of seismic waves, Danish seismologist Inge Lehmann postulates the existence of a solid inner core within the liquid core of the earth.

Russian geochemist D. S. Korzhinskii publishes an extension of thermodynamics that accounts for reactions between rocks and fluids flowing through them.

1937

U.S. physicist Clinton Davisson and English physicist George Paget Thomson share the Nobel Prize for Physics for their experiments on the interaction of light with crystals.

W. L. Bragg's book *Atomic Structure of Minerals* is published.

Finnish chemist and geologist Victor Moritz Goldschmidt tabulates the absolute abundances of chemical elements in Earth from solar and meteorite chemical data.

1938

Seismologists Beno Gutenberg and Charles Richter report the deepest earthquake shock on record; it occurred in 1934 at a depth of 720 kilometers beneath the floor of the Flores Sea, in southern Indonesia.

1939

U.S. chemist A. O. Nier and others report natural variations in the isotopes of carbon. Geochemists will later use the stable isotopes of carbon to study fossil bones, teeth, and rocks.

U.S. chemist Linus Pauling consolidates his theory of the chemical bond in *The Nature of the Chemical Bond, and the Structure of Molecules and*

Crystals, in which he sets out rules for understanding mineral structures. Pauling's rules revolutionize our understanding of the structure and chemistry of minerals.

U.S. geophysicist Walter Maurice Elsasser formulates the dynamo model of the earth, which proposes that eddy currents in the earth's molten iron core cause its magnetism.

1940

Canadian biochemist Martin Kamen discovers carbon 14. It becomes a vital tool in dating young geological samples and archeological remains.

Information from uncrewed weather balloons indicates that columns of warm air rise more than 1.6 km/1 mi above the earth and winds form layers in the lower atmosphere, often blowing in different directions.

1943

Hungarian chemist Georg de Hevesy receives the Nobel Prize for Chemistry for the use of isotopes as tracers.

1946

Carbon 13, a stable isotope of carbon, is discovered. Its abundance will ultimately be used for deciphering the geochemistry of carbon.

U.S. physicist Percy Bridgman receives the Nobel Prize for Physics for his work on high-pressure physics.

1947

English physicist Edward Appleton receives the Nobel Prize for Physics for his discovery of the Appleton layer in the atmosphere.

Russian-born Belgian theoretical chemist Ilya Prigogine publishes theories of irreversible thermodynamics, useful for predicting rates of geochemical processes.

U.S. physicist Willard Libby develops carbon 14 dating.

A University of California, Berkeley, scientist tends to a furnace used in potassium-argon dating to determine the age of materials. (Corbis/Dean Conger)

1948

U.S. chemists L. T. Aldrich and A. O. Nier find argon from decay of potassium in four geologically old minerals, confirming predictions by German physicist C. F. Von Weizsacker made in 1937. The basis for potassium-argon dating is established.

U.S. chemist Harold C. Urey, in a seminal paper, foretells the use of stable isotopes as "geothermometers" and chemical tracers in the earth sciences.

1949

U.S. seismologist Hugo Benioff identifies planes of earthquake foci extending from deep ocean trenches to beneath adjacent continents at approximately 45° as faults. These faults, called Benioff zones or Benioff-Wadati zones, will later be identified as the tops of plates of the lithosphere.

Japanese metamorphic petrologist A. Miyashiro estimates the pressure and temperatures for the aluminosilicate polymorphs (sillimanite, kyanite, and andalusite).

British geophysicist Edward C. Bullard and colleagues design a probe for measuring Earth's heat flow through the ocean floor (published in 1952).

1952

U.S. geophysicist Francis Birch predicts fundamental changes in mineralogy of Earth's mantle with depth.

Radioisotopes begin to be used extensively in scientific research, medicine, and industry. Britain becomes the chief exporter of isotopes.

1953

August 1. The U.S. bathyscaphe *Trieste* is launched. Later in the year it dives to a record 3,150 m/10,300 ft.

U.S. chemists Stanley Miller and Harold Urey demonstrate experimentally how organic compounds might have begun on Earth, by exposing a warm, gaseous mixture of inorganic compounds to electrical discharges. The gases are meant to represent the prebiotic atmosphere, the proverbial primordial soup.

U.S. geophysicist William Ewing announces the existence of a crack, or rift, running along the middle of the Mid-Atlantic Ridge.

U.S. chemist Loring Coes invents an apparatus for obtaining high temperatures at high pressures in the laboratory. He uses this apparatus to grow high-pressure minerals, including a new form of dense silica that bears his name (coesite).

1954

February 15. The bathyscaphe *FNRS 3* descends to 4,000 m/13,000 ft in the Atlantic Ocean off Senegal.

U.S. chemist Linus Pauling receives the Nobel Prize for Chemistry for his discoveries concerning the nature of chemical bonds.

Researchers at General Electric produce the first synthetic diamonds.

The existence of premetazoan life becomes widely accepted when diverse fossil microscopic organisms are discovered in the Gunflint rocks along the north shore of Lake Superior. The Gunflint biota prove that there had been significant evolutionary activity by at least ca. 2.0 billion years before present.

1956

Geochemist C. C. Patterson uses the isotopic composition of lead to determine that three stony meteorites and two iron meteorites have a common age of 4.55 billion years. Patterson uses lead isotopes to establish that Earth is the same age.

U.S. geologists Bruce Heezen and William Ewing discover a global network of oceanic ridges and rifts 60,000 km/37,000 mi long that divide the earth's surface into "plates."

1957

July 1. International Geophysical Year begins. A cooperative international research program, it involves scientists from 70 nations in Antarctic exploration, oceanographic and meteorological research, geomagnetism, seismology, and the launching of satellites into space.

1958

May. Using data from the *Explorer* rockets, U.S. physicist James Van Allen discovers a belt of radiation around the earth. Now known as the Van Allen belts (additional belts were discovered later), they consist of charged particles from the sun trapped by the earth's magnetic field.

The Equatorial Undercurrent is discovered in the equatorial Pacific Ocean. It has a width of 320 –480 km/200 –300 mi, a height of 200 –300 m/650 – 1,000 ft, and flows 50 –150 m/165 –500 ft below the surface.

1959

February 28. The U.S. Air Force launches *Discoverer 1* into a low polar orbit where it photographs the entire surface of the earth every 24 hours. The exposed film is returned to Earth in its ejectable capsule.

U.S. geologists Marion Hubbert and William Rubey demonstrate that the overthrusting of large horizontal slabs of rock that produces folding and mountains is due to the reduction of friction caused by fluids in the rocks.

Russian D. S. Korzhinskii's new rules for the thermodynamics of open systems are published in English in *Physicochemical Basis of the Analysis of the Paragenesis of Minerals* (translated from the Russian of 1957).

1960

January 23. Swiss engineer Jacques Piccard and U.S. Navy lieutenant Don Walsh descend to the bottom of Challenger Deep (10,916 m/35,810 ft), off the Pacific island of Guam, in the bathyscaphe *Trieste,* setting a new undersea record.

The National Center for Atmospheric Research is established at Boulder, Colorado.

U.S. chemist Willard Libby receives the Nobel Prize for Chemistry for his development of radiocarbon dating.

A shot from the back of the National Center for Atmospheric Research in Boulder, Colorado. (Corbis/G. E. Kidder Smith)

U.S. geophysicist Harry Hess develops the theory of seafloor spreading (the term is coined by R. S. Dietz, 1961), in which molten material wells up along the midoceanic ridges, forcing the sea floor to spread out from the ridges. The flow is thought to be the cause of continental drift.

1961

U.S. geophysicist Francis Birch relates seismic velocities to density and average atomic mass of rocks in Earth's mantle.

L. O. Nicolaysen invents the isochron method of rubidium–strontium and uranium–lead dating of geological materials.

U.S. researchers establish Arctic Research Lab Ice Station II, a drifting sea ice station.

1962

U.S. geologist Harry Hess publishes *History of Ocean Basins,* in which he formally proposes seafloor spreading, the idea that the ocean crust is like a giant conveyor belt produced by volcanism at the midoceanic ridges, pushed or pulled away from the ridge axis, and eventually destroyed by plunging down into the deep-sea trenches. Hess's hypothesis, influenced by the earlier work of Arthur Holmes, is embraced by the scientific community.

1963

Canadian geophysicist L. W. Morley discovers the significance of the "striped" pattern of magnetism in rocks of the ocean floor (published in 1964).

British geophysicists Fred Vine and Drummond Matthews analyze the striped pattern of magnetism in rocks of the Atlantic Ocean floor, which assume a magnetization aligned with the earth's magnetic field at the time of their creation. It provides concrete evidence of seafloor spreading.

Geologist I. G. Gass suggests that the Troodos Massif, Cyprus, is a fragment of Mesozoic ocean floor. It is the first recognition of the significance of ophiolites.

1964

The U.S. National Science Foundation establishes a consortium of four leading institutions, called the Joint Oceanographic Institutions for Deep Earth Sampling (JOIDES), for the purpose of drilling into Earth's ocean floor. Results from the project over the next few years will eventually help confirm the theory of seafloor spreading by establishing that the oceanic crust everywhere is less than 200 million years old.

1965

Canadian geologist John Tuzo Wilson publishes *A New Class of Faults and Their Bearing on Continental Drift,* in which he formulates the theory of plate tectonics to explain continental drift and seafloor spreading. He suggests that the midoceanic ridges, deep-sea trenches, and the faults that connect them combine to divide Earth's outer layer into rigid, independent plates. The connecting faults are shown by Wilson to have a unique geometry, and are called transform faults.

French oceanographer Jacques Cousteau heads the Conshelf Saturation Dive Program, which sends six divers 100 m/328 ft down in the Mediterranean for 22 days.

NASA launches *GEOS 1* (Geodynamics Experimental Ocean Satellite). Its aim is to provide a three-dimensional map of the world accurate to within 10 m/30 ft.

The Large Aperture Seismic Array is established in Montana. The signals from 525 seismometers, dispersed over an area of 30,000 sq km/11,600 sq mi, are combined to record seismic events with a high degree of sensitivity.

1966

The National Science Foundation in the United States puts Scripps Institute of Oceanography in charge of the JOIDES project and establishes the Deep Sea Drilling Project, or DSDP.

Geochemists A. E. Ringwood and A. Major relate the seismic discontinuity at a depth of 400 km/250 mi to the change of the mineral olivine to the structure of the mineral spinel.

U.S. geologists Allan Cox, Richard Doell, and Brent Dalrymple publish a chronology of magnetic polarity reversals going back 3 million years. It is useful in dating fossils.

1967

U.S. scientists Syukuvo Manabe and R. T. Wetherald warn that the increase in carbon dioxide in the atmosphere produced by human activities is causing a "greenhouse effect," which will raise atmospheric temperatures and cause a rise in sea levels.

Geophysicist L. R. Sykes uses first-motion seismic studies to establish that midoceanic ridges form with offsets (that is, sections of the ridges are displaced relative to one another along faults so that the crest of the ridge has a stepped appearance) rather than being offset later. This is a major advance in understanding the formation of ocean basins.

British geophysicist D. McKenzie (published with R. L. Parker 1967) and U.S. geophysicist Jason Morgan (published 1968) describe the motions of plates across Earth's surface. Morgan calls the plates the tectosphere. They are later referred to as plates of lithosphere.

1968

French geophysicist Xavier Le Pichon, working at the Lamont Observatory in New York, describes the motions of Earth's six largest plates using poles of rotation derived from the patterns of magnetic anomalies and fracture zones about midoceanic ridges.

British geochemist E. R. Oxburgh and U.S. geodynamicist D. L. Turcotte calculate the thermal consequences of subduction.

A borehole drilled 2,162 m/7,093 ft into the Antarctic ice at Byrd Station reveals that the bottom layers are 100,000 years old.

The U.S. research ship *Glomar Challenger* starts drilling cores in the sea floor as part of the Deep Sea Drilling Project (DSDP). Capable of drilling in water up to 6,000 m/20,000 ft deep, it has technology allowing the return of core samples from 750 m/2,500 ft below the sea floor and is equipped with a gyroscopically controlled roll-neutralizing system that allows it to maintain its stability in diverse weather conditions.

U.S. scientist Elso Sterrenberg Barghorn and his associates report the discovery of the remains of amino acids in rocks 3 billion years old.

1970

British geologist J. F. Dewey, with J. M. Bird, relates the positions of Earth's mountain ranges to the motions of lithospheric plates.

1971

Greenpeace, the environmental campaign organization, is founded to protest against U.S. nuclear testing at Amchitka Island, Alaska.

1972

April 16–27. The United States launches the *Apollo 16* moon mission. Astronauts Charles Duke and John Young return with 97 kg/214 lb of lunar soil and rock after spending a record 71 hours and 2 minutes on the moon.

June. The United Nations Conference on the Human Environment is held in Stockholm, Sweden. The first international conference on the state of the environment, its aim is to improve the world's environment through monitoring, resource management, and education.

July 23. *Landsat 1* is launched, the first of a series of satellites for surveying the earth's resources from space.

The United Nations Environment Program (UNEP) is established; its aim is to advise and coordinate environmental activities within the United Nations.

U.S. paleontologists Stephen Jay Gould and Nils Eldridge propose the punctuated equilibrium model—the idea that evolution progresses in fits and starts rather than at a uniform rate.

1974

Earth scientist John Liu discovers that the lower mantle, comprising most of the earth, is likely composed of silicate perovskite, a mineral with a structure wildly different from the minerals found in Earth's upper mantle and crust.

Mexican chemist Mario Molina and U.S. chemist F. Sherwood Rowland warn that the chlorofluorocarbons (CFCs) used in refrigerators and as aerosol propellants may be damaging the atmosphere's ozone layer, which filters out much of the sun's ultraviolet radiation.

The Global Atmospheric Research Program (GARP) is launched. An international project, its aim is to provide a greater understanding of the mechanisms of the world's weather by using satellites and by developing a mathematical model of the earth's atmosphere.

1975

Five new nations, the USSR, West Germany, France, Japan, and the United Kingdom join the Deep Sea Drilling Project (DSDP) to form the International Phase of Ocean Drilling (IPOD).

David Mao and Peter Bell of the Geophysical Laboratory use a diamond-anvil cell to produce pressures exceeding a million atmospheres.

October. The United States launches the first *Geostationary Operational Environmental Satellite* (GOES); it provides 24-hour coverage of U.S. weather.

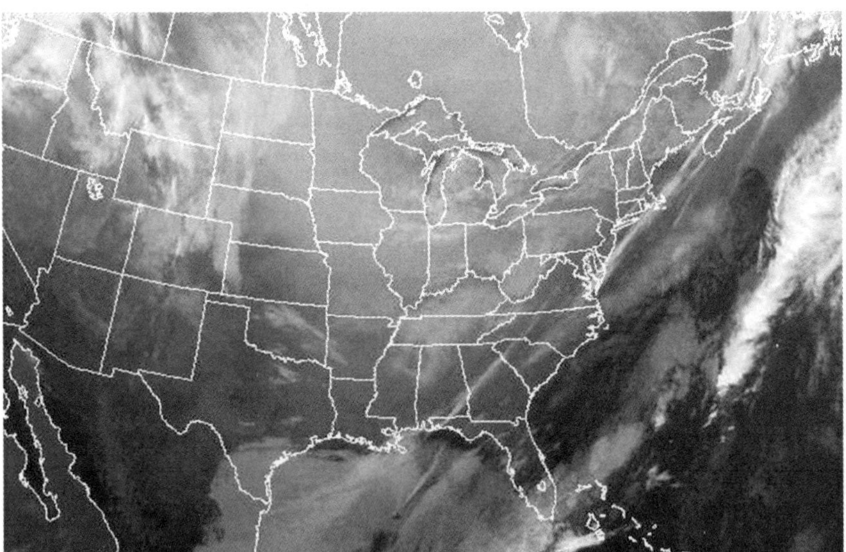

A weather map of the United States as generated by the *Geostationary Operational Environmental Satellite* launched in 1975. (NOAA)

1976

The American Panel on Atmospheric Chemistry warns that the earth's ozone layer may be being destroyed by chloroflurocarbons (CFCs) from spray cans and refrigeration systems.

The U.S. spacecraft *Viking 1* and *Viking 2* (launched in 1975) soft-land on Mars (July 20, September 3). They make meteorological readings of the Martian atmosphere and search for traces of bacterial life. The results prove inconclusive.

The International Magnetosphere Study initiates a three-year observation of the earth's magnetosphere and its effects on the lower stratosphere, including the disruptive effects of magnetic storms.

1977

California enforces strict antipollution legislation, compelling car manufacturers to install catalytic converters that reduce exhaust emissions by 90 percent.

Scientists from the project FAMOUS (French-American Mid-Ocean Undersea Study) in their deep-sea submersible vehicle, *ALVIN,* discover a host of strange life forms, such as large red and white tube worms, near undersea hot springs heated by ocean-ridge volcanism. The discovery proves the existence of life in extreme conditions.

1978

June 27. The U.S. satellite *Seasat 1* is launched to measure the temperature of sea surfaces, wind and wave movements, ocean currents, and icebergs. It operates for 99 days before its power fails.

December. The U.S. space probes *Pioneer Venus 1* (launched May 20) and *2* (launched August 8), go into orbit around Venus, the first relaying information about the atmosphere, the second taking radar photographs of the surface that reveal huge mountains and basins.

Core samples from the seabed are collected by the U.S. research vessel *Glomar Challenger* from a record depth of 7,042 m/23,104 ft.

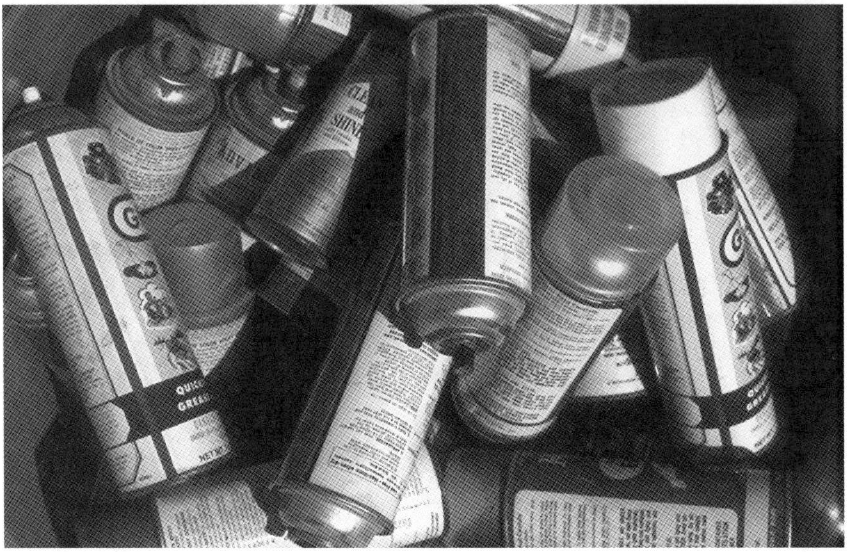

A pile of aerosol cans wait for safe disposal at a Unocal station in Los Angeles, California, in 1990. Aerosol cans pose a serious threat to the environment due to chlorofluorocarbons (CFCs). (Corbis/Joseph Sohm; Chromosohm Inc.)

The United States bans the use of chlorofluorocarbons (CFCs) as spray propellants in order to reduce damage to the ozone layer.

1980

North American geologists P. Coney, D. L. Jones, and J. W. H. Monger describe the North American Cordilleran orogen (a region of mountain-building) in the western United States and Canada as a composite of suspect terranes, resulting in a new perspective on the construction of orogenic ranges (in which mountain ranges eventually form).

May 18. Mount St. Helen's volcano in Washington state erupts explosively in a blast 500 times more powerful than the Hiroshima bomb, causing an outbreak of fires, mudslides, and floods; 34 people die in the largest eruption in U.S. history. Ash from the volcano blankets the surrounding area and affects global temperature readings for months to come.

A thin layer of iridium-rich clay, about 65 million years old, is found around the world. U.S. physicist Luis Alvarez suggests that it was

The eruption of Mount St. Helen's in Washington, 1980. (USGS)

caused by the impact of a large asteroid or comet that threw enough dust into the sky to obscure the sun and cause the extinction of the dinosaurs.

The U.S. *Magsat* satellite completes its mapping of the earth's magnetic field.

1981

The U.S. Committee on the Atmosphere and Biosphere reports evidence linking acid rain to sulfur emissions from power plants.

1982

March–April. El Chichón volcano in Mexico erupts violently, sending dust and gases into the stratosphere.

The Convention on Conservation of Antarctic Marine Living Organisms comes into effect, establishing a protective oceanic zone around the continent.

1983

Research deep-sea drilling ship *Glomar Challenger,* deployed by the Deep Sea Drilling Project (DSDP), is retired. It travelled a total of 375,632 nautical miles and acquired 19,119 cores during its 15-year service.

The Ocean Drilling Program (ODP) is established as the successor to the Deep Sea Drilling Project.

A four-month eruption on Mount Etna in Italy prompts authorities to use dynamite to divert lava flows.

An exceptionally warm El Niño (warm water current) off the coasts of Ecuador and Peru drives the huge schools of anchovies, which thrive in the cold water, further offshore, resulting in the deaths of millions of larger fish and the birds that feed on them and seriously disrupting commercial fishing.

Soviet engineers drill a borehole to a depth of 12.3 km/7.6 mi at Zapolarny in the Kola Peninsula, USSR—the deepest ever drilled.

A fossil land mammal (a marsupial) is discovered in Antarctica.

The skull of a creature called *Pakicetus* is discovered in Pakistan; estimated to be 50 million years old, it is intermediate in evolution between whales and land animals.

U.S. chemist Mark Thiemens and his colleagues demonstrate that the production of ozone in the upper atmosphere causes separation of the two different heavy isotopes of oxygen *independent* of their masses. The discovery of this nonmass-dependent kinetic partitioning of oxygen isotopes provides a new means of tracing the mixing of gases between Earth's different layers of atmosphere.

1984

Australian geologists Bob Pidgeon and Simon Wilde discover zircon crystals in the Jack Hills north of Perth, Australia, that are estimated to be 4.2 billion years old—the oldest minerals ever discovered.

1985

The research drilling ship *SEDCO/BP 471* begins service as part of the new Ocean Drilling Program (ODP), replacing the retired *Glomar Challenger.* The ship is later renamed the *JOIDES Resolution.*

An amphibian skeleton dated 340 million years old is discovered in Scottish oil shale. It is the earliest well-preserved amphibian found.

U.S. chemists Herbert Hauptman and Jerome Karle receive the Nobel Prize for Chemistry for their methods of determining crystal structures.

1987

At a conference in Montreal, Canada, an international agreement, the Montreal Protocol, is reached to limit the use of ozone-depleting chlorofluorocarbons (CFCs) by 50 percent by the end of the century. The agreement is later condemned by environmentalists as "too little, too late."

U.S. researchers prove that thunderstorm systems can propel pollutants into the lower stratosphere when they observe high levels of carbon monoxide and nitric acid at high altitude during a thunderstorm.

1989

1989 is the warmest year on record worldwide; environmentalists suggest this is due to the "greenhouse effect."

The United Nations Environment Program (UNEP) reports that the number of species, and the amount of genetic variation within individual species, is decreasing due to the rapid destruction of natural environments.

1990

Canadian scientists discover fossils of the oldest known multicellular animals, dating from 600 million years ago.

1991

April 4. The U.S. Environmental Protection Agency announces ozone layer depletion at twice the speed previously predicted.

A borehole in the Kola Peninsula in Arctic Russia, begun in the 1970s, reaches a record depth of 12,261 m/40,240 ft.

Less than 50 percent of the world's rainforest remains.

A circular impact structure of Cretaceous–Tertiary (K–T) age is found buried beneath in Mexico's Yucatan Peninsula. Called the Chicxulub crater, it is the best candidate for the K–T impact site envisioned by Alvarez and others.

1992

June 3 –14. The United Nations Conference on Environment and Development is held in Rio de Janeiro, Brazil. It is attended by delegates from 178 countries, most of whom sign binding conventions to combat global warming and to preserve biodiversity (the latter is not signed by the United States).

1993

An ice core drilled in Greenland, providing evidence of climate change over 250,000 years, suggests that sudden fluctuations have been common and that the recent stable climate is unusual.

U.S. scientist Albert Bradley develops the Autonomous Benthic Explorer (ABE), a robotic submersible that can descend to depths of 6.4 km/4 mi and remain at such depths for up to one year.

1995

At the international climate conference held in Melbourne, Australia, scientists report that periodic disruptions of surface currents (which may cause climate changes) have been discovered in the Atlantic and Indian Oceans.

The Prince Gustav Ice Shelf and the northern Larsen Ice Shelf in Antarctica begin to disintegrate—a result of global warming.

U.S. and French geophysicists discover that the Indo-Australian plate split in two in the middle of the Indian Ocean about 8 million years ago.

U.S. chemist F. Sherwood Rowland, Mexican chemist Mario Molina, and Dutch chemist Paul Crutzen receive the Nobel Prize for Chemistry for explaining the chemical processes of the ozone layer.

1996

June 2. U.S. scientists at the National Oceanic and Atmospheric Administration (NOAA) announce the first decline in levels of ozone-depleting chemicals in the air.

Scientists from the Scott Polar Institute, using data from the European Space Agency's *ERS-1* satellite, discover a 125 m-/410 ft-deep lake, 4 km/2.5 mi under the Antarctic ice sheet. Called Lake Vostok after the Russian ice-drilling station it lies beneath, the lake covers 14,000 sq km/5,400 sq mi. The Antarctic ice sheet above, which acts as a blanket, exerts pressures of 300 to 400 atmospheres, allowing the water to remain liquid.

U.S. geophysicists discover that the earth's core spins slightly faster than the rest of the planet.

1997

June 11. French meteorologist Cyril Moulin shows that up to a billion metric tons of dust a year are blown off the arid drought-prone lands surrounding the Sahara Desert in North Africa and carried as far as the

United Kingdom and the Caribbean. The amount has more than doubled in the past 30 years.

June 26. The second Earth Summit takes place in New York. Delegates report on progress since the 1992 Rio Summit and note that progress on the Rio biodiversity convention has been slower than on the convention on climate. The delegates fail to agree on a deal to address the world's escalating environmental crisis. Dramatic declines in aid to the so-called Third-World countries, which the 1992 summit promised to increase, are at the heart of the breakdown.

July 24. Canadian researcher Richard Bottomley and colleagues date the 100-km-/62-mi-wide Popigai impact crater in Siberia, thought to be the fifth largest impact crater on Earth, to 35.7 million years old. They suggest that the meteorite that created it may be responsible for the mass extinction that occurred at the end of the Eocene and the start of the Oligocene geological periods.

July 25. U.S. researcher Joseph L. Kirschvink and colleagues, by examining the record of remnant magnetism in very ancient rocks, suggest that the outer layers of the earth shifted by 90 degrees relative to the core between about 535 and 520 million years ago. This major reorganization of the continents they suggest may have led to the Cambrian Explosion—the rapid appearance of abundant fossils in the geological record in the Cambrian Period, which began 540 million years ago.

August 26. At the World Climate Research Program meeting in Geneva, Switzerland, weather experts predict that El Niño could cause extreme weather conditions during the first six months of 1998 in Asia, Africa, and the United States.

September 7. Australian researcher William de la Mare, using old whaling records which record data on every whale caught since the 1930s, including the ship's latitude, announces the discovery that Antarctic sea ice could have decreased by up to a quarter between the mid-1950s and the 1970s. The finding has major implications, both for global climate conditions as well as for whaling.

Possible fossil evidence for microbial life is discovered in an SNC meteorite widely regarded as having come from Mars; the scientific community intensifies its search for more fossil evidence of extraterrestrial life

Residents of Trujillo, in northern Peru, walk in the Mampuesto cemetery after recent floods due to El Nino caused parts of the terrain to collapse, uncovering buried coffins (12 February 1998). (Corbis/AFP)

and life at extreme conditions. Validity of evidence for past life in the Mars meteorite is questioned immediately.

1998

April 17. An iceberg 40 km/25 mi long and 4.8 km/3 mi wide breaks off from the Larsen B Ice Shelf in Antarctica. Global warming is thought to be the cause.

1999

April. In Price, Utah, paleontologists discover fossils of some of the largest armored dinosaurs ever found. The size of army tanks, ankylosaurs or "fused lizards" are believed to have immigrated to North America from Asia when land masses collided 100 million years ago. In addition, the Price site appears to contain remains of other kinds of dinosaurs previously unknown to paleontologists.

3 Biographical Sketches

Alvarez, Luis Walter (1911–1988)

U.S. physicist. He led the research team that discovered the X_0 subatomic particle in 1959. He also made many other breakthroughs in fundamental physics, accelerators, and radar. He worked on the U.S. atomic bomb for two years, at Chicago and at Los Alamos, New Mexico, during World War II. He was awarded the Nobel Prize for Physics in 1968.

In 1980 Alvarez was responsible for the theory that dinosaurs disappeared because a meteorite crashed into Earth 70 million years ago, producing a dust cloud that blocked out the sun for several years, causing dinosaurs and plants to die.

Alvarez was born in San Francisco and studied at the University of Chicago. During World War II he moved to the Massachusetts Institute of Technology, where he developed the VIXEN radar for the airborne detection of submarines, phased-array radars, and ground-controlled approach radar that enabled aircraft to land in conditions of poor visibility. He also participated in creating the atomic bomb dropped on Hiroshima, Japan. In 1945 he became professor at the University of California, working at the Lawrence Livermore Radiation Laboratory there from 1954 to 1959.

Alvarez built the first practical linear accelerator and an accelerator for breeding plutonium, and invented the tandem electrostatic accelerator. He also devised, but never built, the microtron for accelerating electrons.

In 1953 Alvarez met Donald Glaser, inventor of the bubble chamber detector for subatomic particles. Alvarez decided to build a much larger chamber than Glaser had used, and to fill it with liquid hydrogen. He also developed automatic scanning and measuring equipment, the output of which could be stored on punched cards and then analyzed using computers. Alvarez and his coworkers used the bubble chamber to discover a

Luis Alvarez prepares to evacuate a Geiger counter in his laboratory at the University of California, Berkeley (1946). (Corbis-Bettmann)

large number of new short-lived particles, including the K (the first meson) and the Ω meson. These experimental findings were crucial in the development of the "eightfold way" model of elementary particles and, subsequently, the theory of quarks.

Bjerknes, Vilhelm Firman Koren (1862–1951)

Norwegian scientist whose theory of polar fronts formed the basis of all modern weather forecasting and meteorological studies. He also developed hydrodynamic models of the oceans and the atmosphere and showed how weather prediction could be carried out on a statistical basis, dependent on the use of mathematical models.

Bjerknes was a professor at Stockholm, Sweden, and Leipzig, Germany, before returning to Norway and founding the Bergen Geophysical Institute in 1917.

During World War I, Bjerknes instituted a network of weather stations throughout Norway; coordination of the findings from such stations led him and his coworkers to the theory of polar fronts, based on the discovery that the atmosphere is made up of discrete air masses displaying dissimilar features. He coined the word *front* to denote the boundary between such air masses.

Boltwood, Bertram (1870–1927)

U.S. radiochemist who pioneered the use of radioactive elements as tools for dating rocks and determined the age of the earth to be at least 2.2 billion years old. He discovered ionium and also made the first observations of the phenomenon of isotopy.

Born in Amherst, Massachusetts, Boltwood studied chemistry at Yale from 1889 to 1892 and went on to study at Munich and Leipzig before returning to Yale as a professor from 1897 to 1900. In 1900 he left Yale to work with geologist J. H. Platt. It was during this time that Ernest Rutherford and Frederick Soddy discovered that radioactive elements, such as thorium and uranium, turn into other elements by radioactive decay. By studying the amount of radioactive elements in ores, Boltwood deduced that the radium present in an ore was the product of the breakdown of uranium in the ore and that uranium ultimately would decay to lead.

In 1907 Boltwood demonstrated that by knowing the rate at which uranium decays (its halflife) he could calculate the age of a mineral by measuring the relative proportions of its uranium and lead. Boltwood dated rocks from several localities using his uranium–lead technique, obtaining ages between 410 million to 2.2 billion years old. His efforts showed that the earth was significantly older than previously thought.

Boltwood returned to Yale in 1907 as a professor of radiochemistry. He went on to discover ionium. In his efforts to separate ionium from thorium he made the first observations of isotopes, because ionium is not a separate element, but actually an isotope of thorium.

Bowen, Norman Levi (1887–1956)

Canadian geologist whose work helped found modern petrology (the study of rocks). He demonstrated the principles governing the formation of magma by partial melting, and the fractional crystallization of magma.

Born in Kingston, Ontario, Bowen was educated at the local Queen's University, then moved to the recently founded Geophysical Laboratory in Washington, D.C. His findings on the experimental melting and crystallization behavior of silicates and similar mineral substances were published from 1912 onward. In *The Evolution of Igneous Rocks* (1928) he dealt particularly with magma, becoming known as the head of the "magmatist school" of Canadian geology.

Bridgman, Percy Williams (1882–1961)

U.S. physicist. His research into machinery producing high pressure led in 1955 to the creation of synthetic diamonds by the General Electric company. He was awarded the Nobel Prize for Physics in 1946.

Percy Bridgman at work in his laboratory at Harvard (ca. 1930). (Corbis/Hulton-Deutsch Collection)

Born in Cambridge, Massachusetts, he was educated at Harvard, where he spent his entire academic career.

Bridgman's experimental work on static high pressure began in 1908, and because this field of research had not been explored before, he had to invent much of his own equipment; for example, a seal in which the pressure in the gasket always exceeds that in the pressurized fluid. The result is that the closure is self-sealing. His discoveries included new, high-pressure forms of ice.

His technique for synthesizing diamonds was used to synthesize many more minerals and a new school of geology developed, based on experimental work at high pressure and temperature. Because the pressures and temperatures that Bridgman achieved simulated those deep in the earth, his discoveries gave an insight into the geophysical processes that take place within the earth. His book *Physics of High Pressure* (1931) still remains a basic reference work.

Bullard, Edward Crisp (1907–1980)

English geophysicist who, with U.S. geologist Maurice Ewing, founded the discipline of marine geophysics. He pioneered the application of the seismic method to study the seafloor. He also studied continental drift before the theory became generally accepted.

Bullard was born in Norwich, England, and educated at Cambridge. During World War II he did military research, and he continued to advise the U.K. Ministry of Defense for several years after the war. He was professor of geophysics at the University of Toronto, Canada, from 1948 to 1950; director of the National Physical Laboratory at Teddington, Middlesex, from 1950 to 1957; and head of geodesy and geophysics at Cambridge from 1957 to 1974. He was also a professor at the University of California from 1963 until retirement, and advised the U.S. government on nuclear waste disposal.

Bullard's earliest work was to devise a technique (involving timing the swings of an invariant pendulum) to measure minute gravitational variations in the East African Rift Valley. He then investigated the rate of efflux (outflow) of the earth's interior heat through the land surface; later he devised apparatus for measuring the flow of heat through the deep seafloor.

While at Toronto University, Bullard developed his dynamo theory of geomagnetism, according to which the earth's magnetic field results from convective (circulating) movements of molten material within the earth's core. Bullard was knighted in 1953.

Crutzen, Paul (1933–)

Dutch meteorologist who shared the 1995 Nobel Prize for Chemistry with Mexican chemist Mario Molina and U.S. chemist F. Sherwood Rowland for their work in atmospheric chemistry, particularly concerning the formation and decomposition of ozone. They explained the chemical reactions that are destroying the ozone layer.

While working at Stockholm University in 1970, Crutzen discovered that the nitrogen oxides NO and NO_2 speed up the breakdown of atmospheric ozone into molecular oxygen. These gases are produced in the atmosphere from nitrous oxide N_2O, which is released by microorganisms in the soil. He showed that this process is the main natural method of ozone breakdown. Crutzen also discovered that ozone-depleting chemical reactions occur on the surface of cloud particles in the stratosphere.

Crutzen was born in Amsterdam. He received his doctorate degree in meteorology from Stockholm University in 1973. He is currently at the Max Planck Institute for Chemistry in Mainz, Germany.

Du Toit, Alexander Logie (1878–1948)

South African geologist. His work was to form one of the foundations for the synthesis of the continental drift theory and plate tectonics that created the geological revolution of the 1960s.

The theory of continental drift put forward by German geophysicist Alfred Wegener inspired Du Toit's book *A Geological Comparison of South America and South Africa* (1927), in which he suggested that these continents had probably once been joined. In *Our Wandering Continents* (1937) he maintained that the southern continents had, in earlier times, formed the supercontinent of Gondwanaland, which was distinct from the northern supercontinent of Laurasia.

Du Toit was born near Cape Town and studied there, then went to the United Kingdom and studied at Glasgow and the Royal College of Science, London. He spent 1903 through 1920 mapping for the Geological Commission of the Cape of Good Hope.

Elsasser, Walter Maurice (1904–1991)

German-born U.S. geophysicist. Elsasser pioneered analysis of the earth's former magnetic fields, which are recorded by the magnetic minerals in rocks. His research in the 1940s yielded the dynamo model of Earth's magnetic field. The field is explained in terms of the activity of electric currents flowing in the earth's fluid metallic outer core. The theory premises that these currents are magnified through mechanical motions, rather as currents are sustained in power-station generators.

Born in Mannheim and educated at Göttingen, Elsasser left Germany in 1933 following Hitler's rise to power and spent three years in Paris working on the theory of atomic nuclei. After settling in 1936 in the United States and joining the staff of the California Institute of Technology, he specialized in geophysics. Elsasser became a professor at the University of Pennsylvania in 1947; in 1962 he was made professor of geophysics at Princeton.

Eskola, Pentti Eelis (1883–1964)

Finnish geologist who was one of the first to apply physicochemical postulates on a far-reaching basis to the study of metamorphism, thereby laying the foundations of most subsequent studies in metamorphic petrology.

Throughout his life Eskola was fascinated by the study of metamorphic rocks, taking early interest in the Precambrian rocks of England. Building largely on Scandinavian studies, he was concerned to define the changing pressure and temperature conditions under which metamorphic rocks were formed. His approach enabled comparison of rocks of widely differing compositions in respect of the pressure and temperature under which they had originated.

Eskola was born in Lellainen, Finland, and educated as a chemist at the University of Helsinki before specializing in petrology. In the early 1920s he worked in Norway and in Washington, D.C., United States. He was professor at Helsinki from 1924 until 1953.

Ewing, (William) Maurice (1906–1974)

U.S. geologist whose studies of the ocean floor provided crucial data for the plate tectonics revolution in geology in the 1960s. He demonstrated that midoceanic ridges, with deep central canyons, are common to all oceans.

Using marine sound fixing and ranging seismic techniques and pioneering deep-ocean photography and sampling, Ewing ascertained that the crust of the earth under the ocean is much thinner (5–8 km/3–5 mi thick) than the continental crust (about 40 km/25 mi thick). His studies of ocean sediment showed that its depth increases with distance from the midoceanic ridge, which gave clear support for the hypothesis of seafloor spreading.

Ewing was born in Lockney, Texas, and studied at the Rice Institute in Houston. He developed his geological interests by working for oil companies. In 1944 he joined the Lamont–Doherty Geological Observatory, New York. From 1947 until his retirement, he was professor of geology at Columbia University, while also holding a position at the Woods Hole Oceanographic Institute.

Gardner, Julia Anna (1882–1960)

U.S. geologist and paleontologist whose work was important for petroleum geologists establishing standard stratigraphic sections for rocks in the southern Caribbean.

Gardner's research on the stratigraphic paleontology of the Coastal Plain, Texas, and the Rio Grande Embayment in northeast Mexico led to the publication of *Correlation of the Cenozoic Formations of the Atlantic and Gulf Coastal Plain and the Caribbean Region* (1943, with two coauthors).

Julia Gardner was born in South Dakota and educated at Bryn Mawr College and Johns Hopkins University. She worked for the U.S. Geological Survey from 1911 until 1954. During World War II she joined the Military Geologic Unit where she helped to locate Japanese beaches from which incendiary bombs were being launched by hot air balloons, by identifying shells in the sand ballast of the balloons.

Gibbs, Josiah Willard (1839–1903)

U.S. theoretical physicist and chemist who developed a mathematical approach to thermodynamics and established vector methods in physics. He devised the phase rule and formulated the Gibbs adsorption isotherm.

Gibbs showed how the behavior of materials could be interpreted in terms of the movements of enormous numbers of bodies such as molecules. His ensemble method equated the behavior of a large number of systems at once to that of a single system over a period of time.

Gibbs was born in New Haven, Connecticut, and studied at Yale and in Europe from 1866 to 1869. From 1871 until his death he was professor of mathematical physics at Yale.

The phase rule (published 1876–1878) quantifies the difference between the number of variables and number of governing equations in a thermodynamic system. It may be stated as:

$$f = n + 2 - r$$

where f is the number of degrees of freedom, n the number of chemical components, and r the number of phases—solid, liquid, or gas. Degrees of freedom are the number of quantities such as temperature and pressure that may be altered independent from one another without changing the number of phases.

Gibbs also introduced the concept of free energy as a measure of the likelihood that a physical chemical process will proceed. It is defined in terms of the enthalpy, or heat content, and entropy, a measure of the disorder of a chemical system. From this Gibbs developed the notion of chemical potential, which is a measure of how the free energy of a particular phase depends on changes in composition.

The Gibbs adsorption isotherm showed that changes in the concentration of a component of a solution in contact with a surface occur if there is an alteration in the surface tension.

Goldschmidt, Victor Moritz (1888–1947)

Swiss-born Norwegian chemist. He did fundamental work in geochemistry, particularly on the distribution of elements in the earth's crust. He considered the colossal chemical processes of geological time to be interpretable in terms of the laws of chemical equilibrium.

Using x-ray crystallography Goldschmidt was able to show that, given an electrical balance between positive and negative ions, the most important factor in crystal structure is ionic size. Exhaustive analysis of results from geochemistry, astrophysics, and nuclear physics led to his work on the cosmic abundance of the elements and the links between isotopic stability and abundance. Studies of terrestrial abundance reveal about eight predominant elements. Recalculation of atom and volume percentages lead to the remarkable notion that the earth's crust is composed largely of oxygen anions (90 percent of the volume), with silicon and the common metals filling up the rest of the space.

Goldschmidt was born in Zürich but moved to Norway as a child and studied at the University of Christiania (now Oslo). He was professor and director of the Mineralogical Institute from 1914 to 1929, when he moved to Göttingen, Germany. The rise of Nazism forced him to return to Norway in 1935, but during World War II he had to flee again, first to Sweden and then to Britain, where he worked at Aberdeen and Rothamsted (on soil science). He returned to Norway after the end of the war.

During World War II Goldschmidt carried a cyanide capsule in order to commit suicide should the Germans have invaded Britain. When a colleague asked for one, Goldschmidt told him: "Cyanide is for chemists; you, being a professor of mechanical engineering, will have to use the rope."

Gould, Stephen Jay (1941–)

U.S. paleontologist and writer. In 1972 he proposed the theory of punctuated equilibrium, suggesting that the evolution of species did not occur at a steady rate but could suddenly accelerate, with rapid change occurring over a few hundred thousand years. His books include *Ever Since Darwin* (1977), *The Panda's Thumb* (1980), *The Flamingo's Smile* (1985), *Wonderful Life* (1990), and *Dinosaur in a Haystack* (1997).

Gould was born in New York and studied at Antioch College, Ohio, and Columbia University. He became professor of geology at Harvard in 1973 and was later also given posts in the departments of zoology and the history of science.

Stephen Gould (R) stands with Christian Alfinsen (L) and Francisco Ayala (C) at the National Press Club in Washington. They hold a legal brief attacking Louisiana's "creation-science" statute. All three agree that evolution is the only explanation for the development of man. (Corbis-Bettmann)

Gould has written extensively on several aspects of evolutionary science, in both professional and popular books. His *Ontogeny and Phylogeny* (1977) provided a detailed scholarly analysis of his work on the developmental process of recapitulation. In *Wonderful Life* he drew attention to the diversity of the fossil finds in the Burgess Shale Site in Yoho National Park, Canada, which he interprets as evidence of parallel early evolutionary trends extinguished by chance rather than natural selection.

Gutenberg, Beno (1889–1960)

German-born U.S. geophysicist who determined the depth of Earth's core and contributed to the understanding of Earth's deep interior.

Gutenberg was born in Darmstadt, Germany, and was educated at Darmstadt and Göttingen Universities. He taught at Freiberg University from 1926 until he moved to the California Institute of Technology (Caltech) in the United States in 1930.

As a student in 1914 Gutenberg used velocities of seismic waves to calculate the depth of Earth's core at 2,900 km/1,812 mi. This boundary between Earth's mantle and its core is called the Gutenberg discontinuity.

While at Caltech, Gutenburg made many contributions to our understanding of the structure of Earth's mantle. In 1948 Gutenberg suggested the existence of a low-velocity zone approximately 60–150 km/38–94 mi below the earth's surface in which seismic waves travel more slowly. This zone is now known to be the asthenosphere, the more ductile layer of the mantle on which the earth's lithospheric plates ride.

Hess, Harry Hammond (1906–1969)

U.S. geologist who in 1962 proposed the notion of seafloor spreading. This played a key part in the acceptance of plate tectonics as an explanation of how the earth's crust is formed and moves.

Hess was born in New York and studied at Yale and Princeton, where he eventually became a professor. Beginning in 1931 he carried out geophysical research into the oceans, continuing during World War II while in the navy. Later he was one of the main advocates of the Mohole project, the aim of which was to drill down through the earth's crust to gain access to the upper mantle.

Building on the recognition that certain parts of the ocean floor were anomalously young, and the discovery of the global distribution of midoceanic ridges and central rift valleys, Hess suggested that convection within the earth was continually creating new ocean floor, rising at midoceanic ridges and then flowing horizontally to form new oceanic crust. It would follow that the further from the midoceanic ridge, the older would be the crust—an expectation confirmed by research in 1963.

Hess envisioned that the process of seafloor spreading would continue as far as the continental margins, where the oceanic crust would slide down beneath the lighter continental crust into a subduction zone, the entire operation thus constituting a kind of terrestrial conveyor belt.

Holmes, Arthur (1890–1965)

English geologist who helped develop interest in the theory of continental drift. He also pioneered the use of radioactive decay methods for rock dating, giving the first reliable estimate of the age of the earth.

Holmes was born in Newcastle-upon-Tyne, England, and studied at Imperial College, London. He was appointed in 1924 to be head of the Geology Department at Durham, moving in 1943 to Edinburgh University.

Following the work of Bertram Boltwood, Holmes was convinced that painstaking analysis of the proportions in rock samples of elements formed by radioactive decay, combined with a knowledge of the rates of decay of their parent elements, would yield an absolute age. Beginning in 1913 he used Boltwood's uranium–lead technique systematically to date fossils whose relative (stratigraphical) ages were established, but whose absolute ages were not.

In 1928 Holmes proposed that convection currents within the earth's mantle, driven by radioactive heat, might furnish the mechanism for the continental drift theory broached a few years earlier by German geophysicist Alfred Wegener. In Holmes's view new rocks were forming throughout the ocean ridges. Little attention was given to these ideas until the 1950s.

Holmes's books include *The Age of the Earth* (1913), *Petrographic Methods and Calculations* (1921), and *Principles of Physical Geology* (1944).

Knopf, Eleanora Frances Bliss (1883–1974)

U.S. geologist who studied metamorphic rocks. She introduced the technique of petrofabrics, or structural petrology, to the United States.

Knopf was born in Rosemont, Pennsylvania, and studied at Bryn Mawr College and the University of California at Berkeley. She spent most of her career working for the U.S. Geological Survey. During the 1930s she was also a visiting lecturer at Yale and at Harvard.

In 1913 in Pennsylvania she discovered the mineral glaucophane, previously unsighted in the United States east of the Pacific. In the 1920s Knopf studied the Pennsylvania and Maryland piedmont and the geologically complex mountain region along the New York–Connecticut border.

Petrofabrics, a technique whereby a rock's grain shapes, relationships, and crystallographic orientation are studied on a microscopic scale and used to infer the type of deformation the rock underwent, was developed in Austria at Innsbruck University. Knopf applied this technique for the first time in the United States, publishing the work *Structural Petrology* in 1938.

Le Pichon, Xavier (1937–)

French geophysicist who worked out the motions of Earth's six major lithospheric plates. His work was instrumental in the development of plate tectonics.

Le Pichon was born in Vietnam. He became a research assistant at Columbia University's Lamont–Doherty Geological Observatory in 1963.

In 1968 he published *Sea-Floor Spreading and Continental Drift,* in which he depicted Earth's lithosphere divided into six major plates. The boundaries between the plates were shown to have high seismic activity and occurred along midoceanic ridges, island arcs, active orogenic (mountain-building) belts, and the transform faults revealed earlier by J. Tuzo Wilson.

Le Pichon received many awards, including the Maurice Ewing Medal of the American Geophysical Union (1984) and the Wollaston Medal of the Geological Society of London (1991). He is currently a professor in the Department de Geologie at the Ecole Normale Superieure in Paris, France.

Libby, Willard Frank (1908–1980)

U.S. chemist whose development in 1947 of radiocarbon dating as a means of determining the age of organic or fossilized material won him a Nobel Prize in 1960.

Libby was born in Grand Valley, Colorado, and studied at the University of California, Berkeley. During World War II he worked on the development of the atomic bomb (the Manhattan Project). In 1945 he became professor at the University of Chicago's Institute for Nuclear Studies. He was a member of the U.S. Atomic Energy Commission from 1954 to 1959, and then became director of the Institute of Geophysics at the University of California.

Having worked on the separation of uranium isotopes for producing fissionable uranium-238 for the atomic bomb, he turned his attention to carbon-14, a radioactive isotope that occurs in the tissues of all plants and animals, decaying at a steady rate after their death. He and his co-workers accurately dated ancient Egyptian relics by measuring the amount of radiocarbon they contained, using a sensitive Geiger counter. By 1947 they had developed the technique so that it could date objects, rocks, and fossils up to 50,000 years old.

Matsuyama, Motonori (1884–1958)

Japanese geophysicist who determined that Earth's magnetic field reverses its polarity periodically throughout its geological history. He also pioneered the use of gravimetry in finding geological structures below Earth's surface.

Matsuyama was born in Usa, Japan, and studied mathematics and physics at Hiroshima Normal College and Imperial University, Kyoto. He became a professor at Imperial University in 1913, where he studied gravity determination using pendulum techniques. He carried out extensive gravity surveys in Manchuria, Korea, and in the Japan Trench.

In 1930 Matsuyama began studying magnetism preserved in rocks. It was known as early as 1906 that some rocks were magnetized such that they pointed toward magnetic north, while the magnetism of other rocks pointed in the opposite direction. Matsuyama studied these magnetic anomalies and suggested they were due to a reversal in the polarity of Earth's magnetic field. The Matsuyama Epoch, a major polar reversal occurring approximately .5–2.5 million years ago is named for him.

Miller, Stanley Lloyd (1930–)

U.S. chemist who, in the early 1950s, tried to recreate the formation of life on Earth under laboratory conditions. To water under a gas mixture of methane, ammonia, and hydrogen, he added an electrical discharge. After a week he found that amino acids, the ingredients of proteins—the building blocks of life—had been formed.

Miller was born in Oakland, California, and studied at the universities of California and Chicago. Beginning in 1960 he held appointments at the University of California in San Diego, rising to professor of chemistry.

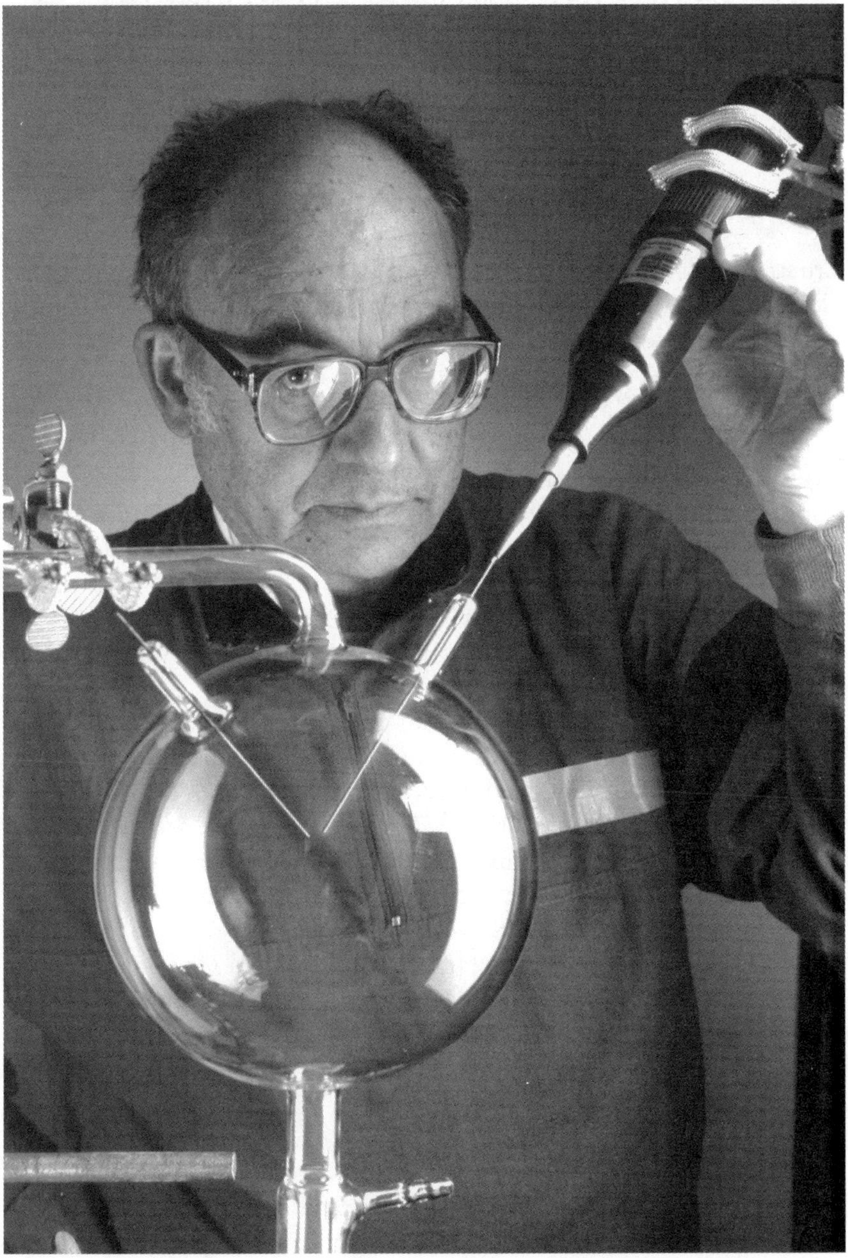

Stanley Miller with the apparatus he developed while he was still a student that simulates conditions of the atmosphere of a newly created Earth (1994). (Corbis/Jim Sugar Photography)

Miller performed his famous experiment while working for his Ph.D. degree under Harold Urey, using the components that had been proposed for the earth's primitive atmosphere by Urey and Russian biochemist Alexandr Oparin. The electrical discharge simulated the likely type of energy source on the primitive earth—a lightning strike.

Mohorovicic, Andrija (1857–1936)

Croatian seismologist and meteorologist who discovered the Mohorovicic discontinuity, the boundary between the earth's crust and the mantle.

Born in Volosko, Croatia, Mohorovicic studied mathematics and physics at Prague University. After seven years as a school teacher, he became a professor at the Zagreb Technical School and later Zagreb University. In 1892 he became director of the Meteorological Station in Zagreb, which later became the Royal Regional Centre for Meteorology and Geodynamics, establishing a seismological observatory there in 1901.

In 1909 after a strong earthquake occurred in the Kulpa Valley south of Zagreb, Mohorovicic discovered two distinct sets of P and S waves—one set arriving earlier than the other. He deduced that one set of waves was slower than the other because it had traveled through denser material.

Mohorovicic surface consists of an outer layer of rocky material approximately 30 km/19 mi thick, which overlies a denser mantle. Later research has shown that the boundary between these two layers, the "Moho," lies at a depth of 5–10 km/3–6 mi beneath the ocean crust and approximately 35 km/21 mi beneath the crust of the continents. Seismic waves travel nearly 20 percent slower below the Moho than above and it is regarded as the bottom of Earth's crust.

The Royal Regional Center for Meteorology and Geodynamics was renamed the Geophysical Institute in 1921. Mohorovicic continued seismological research there until 1926.

Oldham, Richard Dixon (1858–1936)

Irish seismologist who discovered Earth's core and first distinguished between primary and secondary seismic waves.

Oldham was born in Dublin, Ireland, and was educated at the Royal School of Mines. During his tenure at the Geological Survey of India (1879–1903), he investigated an earthquake in Assam in June 1897 and recognized that there were two phases of seismic waves recorded by the seismograph, primary waves and secondary waves, and that these should propagate through the whole earth.

While analyzing seismic records in 1906, Oldham noticed an area on the globe in which P-waves were not detected. Every time an earthquake occurred, this P-wave "shadow zone" appeared on the opposite side of the globe. Oldham demonstrated that the earth had a core that was causing the primary waves to refract (bend) away, leaving a seismic shadow.

Oldham further realized that the material of the core was significantly different from the rest of the earth because the strength of the secondary waves was greatly reduced, creating another shadow zone. In 1919 after the theory of seismic waves had been developed by Wiechert, Oldham suggested that the core may be liquid.

Pauling, Linus Carl (1901–1994)

U.S. theoretical chemist and biologist. His ideas on chemical bonding are fundamental to modern theories of molecular structure and mineralogy. He also investigated the properties and uses of vitamin C as related to human health. He won the Nobel Prize for Chemistry in 1954 and the Nobel Peace Prize in 1962, having campaigned for a nuclear-test ban.

Pauling's work on the nature of the chemical bond included much new information about interatomic distances. Applying his knowledge of molecular structure to proteins in blood, he discovered that many proteins have structures held together with hydrogen bonds, giving them helical shapes.

He was a pioneer in the application of quantum-mechanical principles to the structures of molecules, relating them to interatomic distances and bond angles by x-ray and electron diffraction, magnetic effects, and thermochemical techniques. In 1928 Pauling introduced the concept of hybridization of bonds. This provided a clear, basic insight into the framework structure of all carbon compounds; that is, of the whole of organic chemistry. He also studied electronegativity of atoms and polarization (movement of electrons) in chemical bonds. Electronegativity values can be used to show why certain substances, such as hydrochloric acid, are acid, whereas others, such as sodium hydroxide, are alkaline.

Much of this work was consolidated in his book *The Nature of the Chemical Bond* (1939). Pauling elucidated the nature of the bond between silicon and oxygen, the fundamental building blocks of Earth and the other rocky planets. The majority of rocks are made from silicon chemically bound to oxygen.

Linus Pauling holds up large models of a vitamin C molecule (1981). (Corbis)

In his researches on blood in the 1940s Pauling investigated immunology and sickle-cell anemia. Later work confirmed his conviction that the disease is genetic and that normal hemoglobin and the hemoglobin in sickle cells differ in electrical charge. Pauling's work provided a powerful impetus to Crick and Watson in their search for the structure of DNA.

Pauling was born in Portland, Oregon, and studied at Oregon State Agricultural College, getting his Ph.D. degree from the California Institute of Technology (Caltech). In Europe from 1925 to 1927, he met the chief atomic scientists of the day. He became professor at Caltech 1931, and was director of the Gates and Crellin Laboratories from 1936 to 1958 and of the Linus Pauling Institute of Science and Medicine in Menlo Park, California, from 1973 to 1975.

During the 1950s Pauling became politically active, his special concern being the long-term genetic damage resulting from atmospheric nuclear bomb tests. In this, he came into conflict with the U.S. government establishment and several of his science colleagues. He was denounced as a pacifist and a communist, his passport was withdrawn between 1952 and 1954, and he was obliged to appear before the U.S. Senate Internal Security Committee. One item in his sustained, wide-ranging campaign against nuclear weapons was his book *No More War!* (1958). He presented to the United Nations a petition signed by 11,021 scientists from 49 countries urging an end to nuclear weapons testing, and during the 1960s spent several years on a study of the problems of war and peace at the Center for the Study of Democratic Institutions in Santa Barbara, California.

In the 1970s, Pauling began to argue for the therapeutic value of large doses of vitamin C. In 1971 he wrote *Vitamin C and the Common Cold,* in which he argues that taking vitamin C in daily doses of over 10 grams would also reduce the risk of heart disease. In 1973 he set up the Pauling Institute of Science and Medicine in Palo Alto, California, and for the last few decades of his life took 18 grams of vitamin C a day.

Pauling was coauthor of *Introduction to Quantum Mechanics* (1935); he published two textbooks, *General Chemistry* (1948) and *College Chemistry* (1950).

Thomson, J. J. (Joseph John) (1856–1940)

English physicist who discovered the electron in 1897. His work inaugurated the electrical theory of the atom, and his elucidation of positive rays and their application to an analysis of neon led to the discovery of isotopes. He received the Nobel Prize in 1906 and was knighted in 1908.

Using magnetic and electric fields to deflect positive rays, Thomson found in 1912 that ions of neon gas are deflected by different degrees, indicating that they consist of a mixture of ions with different charge-to-mass ratios. English chemist Frederick Soddy had earlier proposed the existence of isotopes and Thomson proved this idea correct when he identified, also in 1912, the isotope neon-22. This work was continued by his student Frederick Aston.

Thomson was born near Manchester and studied there and at Cambridge, where he spent his entire career. As professor of experimental physics 1884–1918, he developed the Cavendish Laboratory into the world's leading center for subatomic physics. His son was George Paget Thomson, who first demonstrated electron diffraction and the wave nature of the electron.

Investigating cathode rays, Thomson proved that they were particulate and found their charge-to-mass ratio to be constant and with a value nearly 1,000 times smaller than that obtained for hydrogen ions in liquid electrolysis. He also measured the charge of the cathode-ray particles and found it to be the same in the gaseous discharge as in electrolysis. Thus, he demonstrated that cathode rays are fundamental, negatively charged particles; the term *electron* was introduced later.

Urey, Harold Clayton (1893–1981)

U.S. chemist. In 1932 he isolated heavy water and discovered deuterium, for which he was awarded the 1934 Nobel Prize for Chemistry.

Urey was born in Indiana and educated at Montana State University. He became professor of chemistry at Columbia University in 1934, and was at the University of Chicago from 1945 to 1958.

After deuterium, Urey went on to isolate heavy isotopes of carbon, nitrogen, oxygen, and sulfur, ushering in the field of stable isotope geochemistry. His group provided the basic information for the separation of the fissionable isotope uranium-235 from the much more common uranium-238.

Urey also developed theories about the formation of the earth. He thought that the earth had not been molten at the time when its materials accumulated. In 1952 he suggested that molecules found in its primitive atmosphere could have united spontaneously to give rise to life. Along with his student Stanley Miller, Urey performed an experiment to synthesize life from amino acids. The moon, he believed, had a separate origin from the earth.

During World War II Urey was a member of the Manhattan Project, which produced the atomic bomb, and after the war he worked on tritium (another isotope of hydrogen, of mass 3) for use in the hydrogen bomb, but later he advocated nuclear disarmament and world government.

Harold Urey, discoverer of heavy hydrogen (deuterium) points to a map of the moon in his office at the University of Chicago (1950). (Corbis-Bettmann)

Wegener, Alfred Lothar (1880–1930)

German meteorologist and geophysicist whose theory of continental drift, expounded in *Origin of Continents and Oceans* in 1915, was originally known as Wegener's Hypothesis. His ideas can now be explained in terms of plate tectonics, the idea that the earth's surface consists of a number of lithospheric plates, all moving with respect to one another.

Wegener was born in Berlin and studied at Heidelberg, Innsbruck, and Berlin. From 1924 he was professor of meteorology and geophysics at Graz, Austria. He completed three expeditions to Greenland and died while on a fourth.

Wegener supposed that a united supercontinent, Pangaea, had existed in the Mesozoic era. This had developed numerous fractures and had drifted apart some 200 million years ago. During the Cretaceous period, South America and Africa had largely been split, but not until the end of the Quaternary had North America and Europe finally separated; the same was true of the break between South America and Antarctica. Australia had been severed from Antarctica during the Eocene.

Wilson, John Tuzo (1908–1993)

Canadian geologist and geophysicist who established and brought about a general understanding of the concept of plate tectonics.

Born in Ottawa, Wilson studied geology and physics—an original combination that led directly to the development of the science of geophysics—at the University of Toronto. He obtained his doctoral degree at Princeton University, New Jersey.

Wilson's particular interest was the movement of the continents across the earth's surface—then a poorly understood and not widely accepted concept known as continental drift. In *A New Class of Faults and Their Bearing on Continental Drift* (1965), Wilson identified transform faults, which are faults at spreading ridges along which the ridges are offset.

Wilson spent 28 years as a professor of geophysics at the University of Toronto, retiring in 1974—just as interest in plate tectonics was developing worldwide. From then on he was the director-general of the Ontario Science Center and later the chancellor of York University, Toronto, finally retiring in 1987. In 1957 he was the president of the International Union of Geodesy and Geophysics—the most senior administrative post in the field.

Wilson's great strength was in education. He pioneered hands-on interactive museum exhibits, and could explain complex subjects like the movement of continents, the spreading of ocean floors, and the creation of island chains by using astonishingly simple models. He was an active outdoor man, leading expeditions into the remote north of Canada, and he made the first ascent of Mount Hague in Montana, United States, in 1935.

The Wilson Range in Antarctica is named for him.

4 Directory of Organizations

American Association of Petroleum Geologists (AAPG)
1444 South Boulder Avenue
P.O. Box 979
Tulsa, OK 74101
phone: (918) 584-2555
fax: (918) 560-2652
email: postmaster@aapg.org

International organization devoted to advancing the geological sciences as they relate to natural resources, including petroleum, natural gas, and minerals. The principal journal of the AAPG is the *AAPG Bulletin*. They also publish the *AAPG Explorer*, a news magazine, as well as numerous monographs.

American Geological Institute (AGI)
4220 King Street
Alexandria, VA 22302-1502
phone: (703) 379-2480
fax: (703) 379-7563
email: agi@agiweb.org
Web site: http://www.agiweb.org/

Federation of 31 geoscience societies, including the American Geophysical Union (AGU), the American Association of Petroleum Geologists (AAPG), the Geological Society of America (GSA), and many more. Its goal is to provide a united voice for geoscientists in matters of societal concern. AGI publishes a monthly news magazine, *Geotimes*, as well as reference materials such as the *AGI Glossary of Geology* and the GeoRef database for geoscience literature.

American Geophysical Union (AGU)
2000 Florida Avenue NW
Washington, DC 20009-1277
phone: (202) 462-6900 and 1 (800) 966-2481 (toll-free in North America)
fax: (202) 328-0566
email: service@kosmos.agu.org
Web site: http://earth.agu.org/homepage.html

International professional organization for scientists engaged in studies of Earth's oceans, atmospheres, crust, mantle, and core as well as other planets and space. AGU publishes the widely read weekly newspaper *EOS*. It publishes eight scientific journals, including *Geophysical Research Letters, Journal of Geophysical Research, Reviews of Geophysics, Tectonics, Global Biogeochemical Cycles and Nonlinear Processes in Geophysics* (copublished with the European Geophysical Society), *Water Resources Research*, and *Paleoceanography*. In addition, the organization has begun an electronic journal called *Earth Interactions*. There are two well-attended annual AGU meetings each year, the fall meeting in San Francisco and the spring meeting at varied locations in the eastern United States.

American Institute of Hydrology (AIH)
2499 Rice Street, Suite 135
St. Paul, MN 55113-3724
phone: (612) 484-8169
fax: (612) 484-8357
email: aihydro@aol.com or office@aihydro.org
Web site: http://www.aihydro.org/

National organization dedicated to the certification and registration of professionals in the hydrological sciences. The AIH was formed in 1981 as a nonprofit scientific and educational organization.

American Institute of Professional Geologists
7828 Vance Drive #103
Arvada, CO 80003-2125
phone: (303) 431-0831
fax: (303) 431-1332
email: aipg@aipg.com

International association that certifies professional geologists and promotes their interests by lobbying on their behalf at the state and federal levels and working with regulatory boards and agencies.

Association of Engineering Geologists (AEG)
323 Boston Post Road, Suite 2D
Sudbury, MA 01776
phone: (978) 443-4639
fax: (978) 443-2948
email: aeghq@aol.com
Web site: http://geoweb.tamu.edu/aeg/Root_files/

National organization promoting the application of geological techniques to construction and engineering. The AEG has several publications, including the *Environmental & Engineering Geoscience Journal.*

Bavarian Research Institute of Experimental Geochemistry and Geophysics
Bayerisches Geoinstitut
Universität Bayreuth
D-95440 Bayreuth
Germany
phone: +49(0)921 55-3700/3766
fax: +49(0)921 55-3769
email: Bayerisches.Geoinstitut@uni-bayreuth.de
Web site: http://www.bgi.uni-bayreuth.de/

Highly regarded institute at the University of Bayreuth in Germany focusing on high-temperature, high-pressure experimental research in mineralogy, petrology, geochemistry, and geophysics. Specifically, the institute seeks an understanding of Earth's deep interior and its structure and composition.

Current areas of research include the dynamics of Earth's core, the processes of subduction, the formation of the core and mantle in Earth's early evolutionary history, volcanism and its effects on the atmosphere, and the transport of heavy metals in supercritical fluids.

Carnegie Institution of Washington
1530 P Street NW
Washington, DC 20005
phone: (202) 387-6400
Web site: http://www.ciw.edu/ *or* http://www.ciw.edu/Geo_seismo.html

Nonprofit organization founded and funded by Andrew Carnegie and incorporated by an act of Congress in 1904. Carnegie, who had more money than the U.S. Treasury at the time, endowed the institution with $10 million, envisioning a research facility that would "encourage, in the broadest and most liberal manner, investigation, research, and discovery, and the application of knowledge to the improvement of mankind."

Research at the institute includes the earth sciences, astronomy, and biology, with an emphasis on predoctoral and postdoctoral education. As well as its introductory Web site, the institute also maintains a web listing of links to seismology, volcano, and earthquake sites, as well as links to major institutions and universities.

Center for Earth and Planetary Studies (CEPS)
Regional Planetary Image Facility
National Air and Space Museum, MRC 315
Smithsonian Institution
Washington, DC 20560
phone: (202) 357-1457
fax: (202) 786-2566
Web site: http://www.nasm.edu/ceps/rpif.html

A special unit within the Collections and Research Department of the National Air and Space Museum of the Smithsonian Institution, CEPS conducts research in the planetary sciences, terrestrial geophysics, and remote sensing. Additionally it is responsible for two exhibits in the National Air and Space Museum: the Exploring the Planets gallery and the Looking at Earth gallery, which focuses on the use of aerial photography and satellite imaging of the Earth. CEPS is also responsible for all lunar rocks at the museum.

Remote sensing is the technique of using reflected and/or radiated electromagnetic radiation to study the earth. The techniques involve cameras, microwave frequency receivers, and infrared detectors. CEPS seeks an understanding of the geology of our planet through the remote-sensing data obtained by satellites and space missions.

The center is also involved in educational programs, conducting seminars and courses, and is in the process of making remote-sensing data available on the Internet.

Conservation Analytical Laboratory (CAL)
Smithsonian Institution, MSC, MRC 534
Washington, DC 20560
or
Conservation Analytical Laboratory
Smithsonian Institution, MSC D2002
4210 Silver Hill Road
Suitland, MD 20746

phone: (301) 238-3700
fax: (301) 238-3709
email: cal.web@cal.si.edu.

Part of the Smithsonian Institution dedicated to the preservation and conservation of artifacts and their constituent materials in the museum collections. A highly specialized facility, the laboratory's research includes the areas of chemistry, materials technology, treatment processes, art and cultural history, biology, and molecular biology. In addition to research, CAL offers courses on such topics as the preservation of statues, and the use of partly degraded DNA and protein to determine an animal's diet and behavior in the past.

Department of Earth Sciences, The Open University
Walton Hall
Milton Keynes, MK7 6AA
United Kingdom
tel. +44 1908 653012
fax.: +44 1908 655151
Web site: http://exodus.open.ac.uk/

Research at the Open University includes the Petrogenesis Research Group (PRG), a Remote Sensing/Image Processing Unit, an Electron Microprobe Laboratory, and Argon Laser Laboratory. The department has links to the university's Knowledge Media Institute, seeking to use the latest media technologies to convey information in the earth sciences. "The Virtual Earth" is an example of such an endeavor, providing sites such as Virtual Field Trips, Virtual Microscope, and Interacting with Minerals.

Department of Earth Sciences, University of Cambridge
Downing Street
Cambridge, CB2 3EQ
United Kingdom
phone: +44 1223 333400
fax: +44 1223 333450
Web site: http://www.esc.cam.ac.uk/
or
Bullard Laboratories
Madingley Road
Cambridge CB3 OEZ
United Kingdom
tel. +44 1223 337191

Comprised of 32 academic staff members and 40 senior research staff members and fellows, this large department emphasizes its commitment to research. Research departments include: Magmatic and Metamorphic Processes, Geophysics, Institute of Theoretical Geophysics, Mineral Sciences, Sediments and Basins, Geochemistry, Geology, Palaeontology, and the Godwin Institute for Quaternary Research.

Department of Earth Sciences, University of Manchester
Oxford Road
Manchester M13 9PL
United Kingdom
phone: +44 161 275 3804
fax: +44 161 275 3947
email: adminw@fs1.ge.man.ac.uk
or
Geoscience Research Institute
University of Manchester
Oxford Road
Manchester M13 9PL
United Kingdom
phone: +44 161 275 3801

The study of geology at Manchester began in 1851, concentrated on the areas of paleontology and stratigraphy. In the 1950s the department established the first laboratories for experimental petrology in the United Kingdom. It is now recognized as one of the six major centers in the country for earth science teaching and research (along with such institutions as Oxford and Cambridge). The department is a partner in the Greater Manchester Geological Unit, establishing an extensive database and offering technical expertise as well as surveying, drilling, mapping, and aerial photography. It has also benefited from its involvement in the university's interdisciplinary Geoscience Research Institute, which includes faculty from the departments of geology, physics, chemistry, metallurgy, environmental biology, and geography in the university and from the department of chemistry and the Corrosion and Protection Centre in University of Manchester Institute of Science and Technology (UMIST). Along with industry advisers, the research in the project is administrated by Professor C. D. Curtis of Petroleum Science, Professor D. J. Vaughan (chairman of the board) of Ore and Industrial Mineral Deposits, Professor C. M. B. Henderson of Mineral Sciences, and Professor G. Turner of Isotope Geochemistry and Cosmochemistry.

Department of Earth Sciences, University of Oxford
Parks Road
Oxford OX1 3PR
United Kingdom
phone: +44 1865 272000
fax: +44 1865 272072
email: enquiries@earth.ox.ac.uk
Web site: http://www.earth.ox.ac.uk

Regarded as the birthplace of earth sciences in the United Kingdom, Oxford's department of Earth Sciences dates back to the geology lectures given in 1819 by renowned geologist William Buckland, the department's first professor. The department boasts of famous students like Charles Lyell. Its associated University Museum, opened in 1860 to accommodate its natural history collections, was the site of the famous historic debate at the British Association meeting in which Bishop Wilberforce of Oxford debated Thomas Henry Huxley over Darwin's recently published *Origin of Species.*

The Earth Sciences Department is one of the leading centers for geological research in the United Kingdom, with principal areas in geodesy and geodynamics, seismology, geochemistry (isotope geochemistry), mineral physics and chemistry, petrology, tectonics, sedimentology and basin analysis, stratigraphy, paleobiology, palacomagnetism and rock magnetism, and marine geology.

Department of Geological Sciences,
University of South Carolina
700 Sumter Street
Columbia, SC 29208
phone: (803) 777-4535
fax: (803) 777-6610
Web site: http://www.geol.sc.edu/
or
Marine Science Program
University of South Carolina
College of Science and Mathematics
Columbia, SC 29208
phone: (803) 777-2692
Web site: http://inlet.geol.sc.edu/marine-science/

With 23 faculty members in environmental geosciences, geochemistry, geophysics, sedimentology, structural geology, and tectonics, this department is primarily known for its exceptional program in marine sciences. In addition to marine sciences, there are three other primary areas of study emphasized in the department: environmental geosciences, evolution of orogenic systems, and global climate change.

The Marine Science Program at the University of South Carolina is ranked fifth out of 115 marine programs in the United States, providing a wide range of undergraduate and graduate research and education. Research areas include marine ecology; geochemistry; evolution; resource management; ecological, physiological, and population genetics; physical oceanography; seafloor metamorphosis; atmosphere and open ocean dynamics; micropaleontology; sedimentology; microbial processes; and pollution chemistry. The faculty reflects the truly multidisciplinary nature of marine science, with members from the Colleges of Science and Mathematics, Biological Sciences, Chemistry, Biochemistry, Geological Sciences, and also from the Department of Environmental Health Sciences in the School of Public Health.

Department of Geology and Geophysics, University of Edinburgh
Grant Institute
West Mains Road
Edinburgh EH9 3JW
Scotland
phone: +44 131 650 4843
fax: +44 131 668 3184
Web site: http://www.glg.ed.ac.uk/admin/dptcntct.html

The Department of Geology and Geophysics is one of the largest Earth Science departments in the United Kingdom. Principal research areas are igneous, metamorphic, and experimental petrology; tectonics and earth history; marine and quaternary geology; geophysics; and petroleum geoscience.

Department of Geophysical Sciences, University of Chicago
Henry Hinds Laboratory
5734 South Ellis Avenue
Chicago, IL 60637
tel. (773) 702-8101
fax.(773) 702-9505
Web site: http://geosci.uchicago.edu/

One of the most highly regarded earth science departments in the United States, the Department of Geophysical Sciences at the University of Chicago conducts research in the areas of climate and global change, atmospheric dynamics, high pressure geophysics and geochemistry, environmental chemistry, igneous and metamorphic petrology, and paleobiology. The department also includes the Paleogeographic Atlas Project and is well known for its research in geochemistry and cosmochemistry.

Department of Terrestrial Magnetism (DTM), Carnegie Institution of Washington
5241 Broad Branch Road NW
Washington, DC 20015-1305
phone: (202) 686-4370
Web site: http://www.ciw.edu/DTM.html

Part of the Carnegie Institution of Washington, DTM's original purpose was to map the earth's magnetic field. Its research has expanded to include geophysics, geochemistry, cosmochemistry, astrophysics, and planetary physics, with the goal of understanding "the physical Earth and the universe that is our home."

Geochemisches Institut, Göttingen
Goldschmidtstr. 1
37077 Göttingen
Germany
phone: +49 551 39-3972 or +49 551 39-3952
fax. +49 551 39-3982
Web site: http://www.uni-geochem.gwdg.de/docs/home.htm.

One of the institutes in the Department of Earth Science at the University of Göttingen in Germany. The university itself was founded in 1737 by Georg August, who was at that time elector in Hanover and reigned in London as King George II. Notable twentieth-century earth scientist Victor M. Goldschmidt, who came to Göttingen to chair the mineralogy and petrology department, greatly advanced the field of geochemistry at Göttingen, adding a theoretical perspective to the discipline. Goldschmidt sought the principles and processes governing the distribution of elements in rocks and minerals in earth systems and consequently in the "total earth as a cosmic product."

Geological Society of America (GSA)
3300 Penrose Place
P.O. Box 9140
Boulder, CO 80301-9140
phone: (303) 447-2020
fax: (303) 447-1133
email: web@geosociety.org
Web site: http://www.geosociety.org/

American professional organization for earth scientists from all sectors, including academia, government, and industry. The GSA caters to the geological sciences and publishes two major journals, *Geology* and the *GSA Bulletin*. In addition to a popular annual meeting, the society sponsors regional meetings within the United States. The GSA also sponsors several Penrose Conferences each year, with the aim of fostering free exchange of scientific ideas on specified topics.

Geophysical Laboratory, Carnegie Institution of Washington
5251 Broad Branch Road, NW
Washington, DC 20015-2410
phone: (202) 686-2410
fax: (202) 686-2419
Web site: http://gl.ciw.edu/

Part of the Carnegie Institution of Washington, the Geophysical Laboratory was founded in 1905 to understand the physiochemical processes that govern the formation of Earth's crust, mantle, and core.

Geoscience Information Society
c/o American Geological Institute
4220 King Street
Alexandria, VA 22302
Web site: http://www.geoinfo.org/

Society of librarians, scientists, educators, and editors promoting the exchange of information in the geosciences. The society publishes two major reference books: *The Union List of Geologic Field Trip Guidebooks of North America* (6th ed., 1996) and the *Directory of Geosciences Libraries, U.S. and Canada* (5th ed., 1997).

IUCN/World Conservation Union
Rue Mauverney 28
CH-1196 Gland

Vaud
Switzerland
phone: +41 22 999 0001
fax: +41 22 999 0002
email: eld@hq.iucn.org
Web site: http://www.iucn.org/

International conservation organization established in 1948. It links governments and governmental and nongovernmental agencies with scientists and experts to encourage a worldwide approach to conservation. IUCN (International Union for Conservation of Nature) works with six global commissions formed to assist in the protection of nature in the following areas: species survival, protected areas, education and communication, environmental law, ecosystem management, and environmental economics and social policy. IUCN publications include the *Red List of Threatened Animals* and the *Red Data Books.*

Joint Oceanographic Institutions, Inc. (JOI)
Ocean Drilling Program
1000 Discovery Drive
Texas A & M University
College Station, TX 77845-9547
phone: (409) 845-2673
email: Webmaster@odp.tamu.edu
Web site: http://www.joi-odp.org/JOI/JOIHime.html

JOI manages the Ocean Drilling Program (ODP), an international venture in geological oceanography consisting of the efforts of and financial support from the United States (specifically the National Science Foundations), Germany, France, Japan, and the United Kingdom, in full partnership. Australia, Canada, Chinese Taipei, and Korea hold a joint partnership (PacRim). Other partners include China, and the European Science Foundation, consisting of Belgium, Denmark, Finland, Iceland, Italy, the Netherlands, Norway, Portugal, Spain, Sweden, Switzerland, and Turkey. The program conducts research into the evolution of Earth's ocean basins and their underlying crust and is managed by ten U.S. institutions, including Texas A & M University and the Lamont-Doherty Earth Observatory (LDEO) of Columbia University.

The ODP evolved from the Deep Sea Drilling Project (DSDP), overseen by Scripps Institute of Oceanography in 1968 after the formation of JOIDES (the Joint Oceanographic Institutions for Deep Earth Sampling) in 1964. The original drilling vessel was the *Glomar Challenger.*

In 1983, the National Science Foundation designated Texas A&M University as the science operator. At this time, the name Ocean Drilling Program was adopted.

The current drilling vessel, *JOIDES Resolution*, is the best-equipped marine research vessel in the world, with laboratories for downhole measurements; core handling, sampling, and description; physical properties; paleomagnetism; paleontology; thin-section preparation; chemistry; x-ray analysis; and photography.

Lamont-Doherty Earth Observatory (LDEO)
P.O. Box 1000
61 Route 9W
Palisades, NY 10964-1000
phone: (914) 359-2900
fax: (914) 359-2931
email: director@ldeo.columbia.edu
Web site: http://www.ldeo.columbia.edu/

Part of Columbia University, the observatory was founded in 1949 by geologist Maurice Ewing and is located north of New York City on a 125-acre estate donated by Thomas W. Lamont. Lamont was instrumental in the development of plate tectonics and seismic instrumentation, and developing the first global seismic network. Today, research at Lamont is often interdisciplinary, involving seismology, marine geology, marine geophysics, geochemistry, terrestrial geology, climate and atmospheric science, and paleontology and micropaleontology.

Two affiliated institutions, the American Museum of Natural History and the NASA Goddard Institute for Space Studies, collaborate with Lamont researchers.

Lawrence Livermore National Laboratory, University of California
7000 East Avenue
P.O. Box 808
Livermore, CA 94550
phone: (510) 422-1100
fax. (510) 422-1370
Web site: http://www.llnl.gov/

Lawrence Livermore is one of the largest geoscience research institutes in the United States, conducting basic research in the areas of energy, defense, environment, and geosciences. The lab was established in 1952 by the federal Atomic Energy Commission at the instigation of atomic physicist Edward Teller to be a second atomic weapons laboratory, in addition to Los Alamos.

Earth sciences at Livermore is conducted by two programs, the Geosciences and Environmental Technology (GET) program and the Geophysics and Global Security (GGS) program. There are approximately 100 researchers from a range of disciplines, including seismology, physics, geophysics, geochemistry, fluid dynamics, hydrology, material science, chemical and mechanical engineering, and geology.

In turn, GET and GGS host the Center for Geosciences of the Institute for Geophysics and Planetary Physics (IGPP), which collaborates with researchers and faculty members of the University of California campuses. Several technical support groups—Computations, Electrical Engineering, Mechanical Engineering, Chemistry and Materials Science, and the Technical Information Department—add a further 60 scientists and technicians to Livermore's Center for Geosciences complex. Basic geoscience research includes high-pressure geophysics, tectonics, and the properties (chemical, physical, and kinetic) of geological and nongeological materials.

Livermore's scientists and engineers conduct experimental and theoretical modeling studies of the environment. The toxic waste disposal issue is one example of how Lawrence Livermore's multidisciplinary expertise works to resolve an environmental problem. Scientists are studying the geochemistry, hydrology, thermomechanical response, and radionuclide transport at a proposed high-level nuclear waste repository at Yucca Mountain, Nevada.

Energy research at Lawrence Livermore includes "geophysical and seismological studies of geothermal fields, studies of the chemical kinetics of petroleum formation, and seismic, geochemical, and geophysical studies related to oil and gas exploration and production."

Mineralogical Society of America
1015 Eighteenth Street NW, Suite 601
Washington, D.C. 20036
phone: (202) 775-4344
fax: (202) 775-0018
email: business@minsocam.org
Web site: http://www.minsocam.org/

International organization devoted to advances in mineralogy, crystallography, and petrology founded in 1919. The society publishes the journal *American Mineralogist* and a topical monograph series called

Reviews in Mineralogy, which is popular worldwide. It also publishes a newsletter, *The Lattice.*

National Geophysical Data Center, National Oceanic and Atmospheric Administration (NOAA)
Mail Code E/GC
325 Broadway
Boulder, CO 80303-3328
phone: (303) 497-6826
fax: (303) 497-6513
email: info@ngdc.noaa.gov
Web site: http://www.ngdc.noaa.gov/mgg/mggd.html

Part of the U.S. Department of Commerce's National Oceanic and Atmospheric Administration (NOAA) that coordinates and manages environmental data from its five centers: marine geology and geophysics, paleoclimatology, solar-terrestrial physics, solid earth geophysics, and glaciology (snow and ice). As well as fostering research in these areas, the center aims to make data from its 300-plus databases accessible to a variety of users.

National Museum of Natural History
Tenth Street and Constitution Avenue NW
Washington, DC 20560
phone: (202) 357-1729
Web site: http://www.mnh.si.edu/nmnhweb.html

Part of the Smithsonian Institution, the museum was built in 1910 and now covers a total floor area of 20 acres. The museum's focus is on the natural history of humankind and evolution of our natural world. Popular exhibits include dinosaurs, mammals, minerals, an insect zoo, and the Discovery Room. The museum is also home to the legendary Hope Diamond and an eight-ton African bush elephant, which is prominently displayed in the Rotunda.

Paleontological Research Institution
1259 Trumansburg Road
Ithaca, NY 14850
phone: (607) 273-6623
fax: (607) 273-6620
Web site: http://www.englib.cornell.edu/pri/

Institution and museum affiliated with Cornell University that fosters research in paleontology and houses extensive fossil collections, as well as laboratories, offices, and a 50,000-volume research library. The institute publishes *Bulletins of American Paleontology*, billed as "the oldest continuously published paleontological journal in the western hemisphere and one of the oldest in the world."

Paleontological Society, Inc.
P.O. Box 1897
Lawrence, KS 66044-8897
Web site: http://www.uic.edu/orgs/paleo/homepage.html

Society founded in 1908 to promote advances in paleontology. Its members include professional scientists, land managers, museum specialists, and technicians, as well as amateur collectors. The Society publishes the bimonthly *Journal of Paleontology* and *Paleobiology*, which is published quarterly.

Scientific Committee on Problems of the Environment (SCOPE)
51 Boulevard de Montmorency
75016 Paris
France
phone: +33 1 4525 0498
fax: +33 1 4288 1466
email: scope@paris7.jussieu.fr
Web site: http://www.lmcp.jussieu.fr/icsu/Structure/scope.html

International organization formed in 1969 to provide information to organizations and agencies involved in the study of the environment. To this end, SCOPE reviews and assesses changes to the environment, as well as evaluates the methods of measurement used.

Scripps Institution of Oceanography, University of California, San Diego
9500 Gilman Drive
La Jolla, CA 92093-0210
phone: (619) 534-2830
fax: (619) 534-5306
email: siocomm@sio.ucsd.edu
Web site: http://www@sio.ucsd.edu/

Founded originally as the Marine Biological Association of San Diego by William E. Ritter in 1903, the institute was later endowed by Ellen Browning Scripps and her brother and became part of the University of California in 1912. It received its formal name in 1925. Scripps supports laboratory and field research into all aspects of Earth's oceans; geology, geophysics, physics, and chemistry. Its research programs include studies of the marine food chain, earthquakes, coastal processes, climate change, even the possibility of obtaining pharmaceuticals from ocean life.

Seismological Society of America
201 Plaza Professional Building
El Cerrito, CA 94530
phone: (510) 525-5474
fax: (510) 525-7204
email: info@seismosoc.org
Web site: http://www.seismosoc.org

Organization of geologists, geophysicists, seismologists, engineers, and policy makers promoting developments in earthquake science. It publishes the *Seismological Research Letters* and the *Bulletin of the Seismological Society of America*, which claims to be "the premier journal of research in earthquake seismology and related disciplines since 1911."

Smithsonian Environmental Research Center (SERC)
P.O. Box 28
Edgewater, MD 21037-0028
phone: (301) 261-4084
fax: (301) 261-7954
Web site: http://www.serc.si.edu/

One of the research institutes of the Smithsonian Institution that conducts interdisciplinary studies of interactions within ecosystems to improve our "stewardship of the biosphere." Located on the Chesapeake Bay near Annapolis, Maryland, SERC studies the relationships of atmospheric, terrestrial, and aquatic environments. SERC's locality consists of 2,500 acres, including 12 miles of shoreline on the Rhode River subestuary of the Chesapeake Bay. Ecosystems include wetlands and marshes, coastal plain forests, and open estuary waters, as well as agricultural fields, giving scientists the opportunity to study the effects of pollution, fishing, and development.

The Smithsonian Environmental Research Center also has opportunities for interns at the undergraduate or graduate level to conduct research projects supervised by the center's research scientists. Projects include estuarine or terrestrial environmental research, resource planning, and environmental education research and development.

Society for Mining, Metallurgy, and Exploration, Inc.
8307 Shaffer Parkway
P.O. Box 625002
Littleton, CO 80162-5002
phone: (303) 973-9550 X210
fax: (303) 973-3845
email: smenet@aol.com
Web site http://www.smenet.org

International organization established in 1857 to promote mining and minerals exploration. Its publications include *Mining Engineering* (monthly) and *Minerals and Metallurgical Processing* (quarterly).

Society for Organic Petrology
Kentucky Geological Survey
228 Mining and Minerals Building
University of Kentucky
Lexington, KY 40506
phone: (606) 257-5500
fax: (606) 258-1049
email: eble@kgs.mm.uky.edu
Web site: http://www.tsop.org

Organization devoted to studies of the origin of coal and kerogen, and organic geochemistry in general.

Society for Sedimentary Geology (SEPM)
1731 East 71st Street
Tulsa, OK 74136-5108
phone: (918) 493-3361 or (800) 865-9765
fax: (918) 493-2093
email: ngeeslin@sepm.org
Web site: http://www.sepm.org/sepm.HTML

Society dedicated to the dissemination of scientific information on sedimentology, stratigraphy, paleontology, environmental sciences, marine geology, and hydrogeology. The society publishes the *Journal of Sedimentary Research* and *PALAIOS*.

Soil Science Society of America
677 South Segoe Road
Madison, WI 53711
phone: (608) 273-8080
fax: (608) 273-2021
Web site: http://www.soils.org

Promotes advancements in soil research. Its publications include the *Soil Science Society of America Journal*, the *Journal of Environmental Quality*, and the *Journal of Production Agriculture*. Date established, 1936.

United Nations Educational, Scientific and Cultural Organization (UNESCO)
7 Place de Fontenoy
75352 Paris 07 SP
France
phone: (33 1) 4568 1000
fax: (33 1) 4567 1690
Web site: http://www.unesco.org/

International organization established in 1945. Its aim is to contribute to world security and peace through education, science, culture, and communication among nations. UNESCO performs five main functions toward this aim: conducting prospective studies, sharing knowledge, setting standards, lending expertise, and facilitating exchange of specialized information. It currently has 186 member states.

United States Geological Survey (USGS)
Eastern Region and Headquarters:
USGS National Center
12201 Sunrise Valley Drive
Reston, VA 20192
phone: (703) 648-4000
Web site: http://www.usgs.gov/major_sites.html
Central Region:
U.S. Geological Survey
P.O. Box 25046 Denver Federal Center
Denver, CO 80225
phone: (303) 236-5900
Western Region:
U.S. Geological Survey

345 Middlefield Road
Menlo Park, CA 94025
phone: (650) 853-8300
Web site: http://www.usgs.gov/network/science/earth/earthquake.html

Part of the U.S. Department of the Interior, the USGS was established in 1879 under the presidency of Rutherford B. Hayes for the purpose of "classification of the public lands, and examination of the geological structure, mineral resources, and products of the national domain." Its divisions include a biological resources division, geologic division, national mapping division, and a water resources division.

World Meteorological Organization
41, avenue Giuseppe-Motta
1211 Geneva 2
Switzerland
phone: (041 22) 730 8314/15
fax: (041 22) 733 2829
Web site: http://www.wmo.ch

An agency, part of the United Nations since 1950, that promotes the international exchange of weather information through the establishment of a worldwide network of meteorological stations. It was founded as the International Meteorological Organization in 1873, and its headquarters are now in Geneva, Switzerland.

5 Selected Works for Further Reading

Barry, Roger G., and Richard J. Chorley. *Atmosphere, Weather and Climate,* 6th ed. London, New York: Routledge, 1992. 392 pages. ISBN: 0415077605.

A heavyweight volume for the dedicated reader. Everything is here, from the gas content of the sky to the causes of climatic change.

Begon, M., J. L. Harper, and C. R. Townsend. *Ecology: Individuals, Populations, and Communities,* 2d ed. Boston: Blackwell Scientific Publications, 1990. 945 pages. ISBN: 0865421110.

One of the more readable of a series of textbooks on ecology, looking at relationships within populations and communities.

Briggs, David. *Fundamentals of the Physical Environment.* London, New York: Routledge, 1997. 557 pages. ISBN: 041508394X.

Solid and dependable textbook.

Brown, G. C., et al. *Understanding the Earth.* Cambridge, New York: Cambridge University Press, 1992. 551 pages. ISBN: 0521370205.

The updated version of the British Open University's textbook on earth science, bringing the original 1972 classic completely up-to-date.

Budiansky, Stephen. *Nature's Keepers: The New Science of Nature Management.* New York: Free Press, 1995. 310 pages. ISBN: 0029049156.

An environmental journalist explains in simple terms the new science of mathematical ecology, which is providing tools for effective environmental management by revealing how ecosystems really work and interact.

Calder, Nigel. *Restless Earth: A Report on the New Geology.* New York: Viking Press, 1972. 152 pages. ISBN: 0670595306.

This book was landmark in being one of the first general readers (it accompanied a television series) to examine the role of the newly forming concept of plate tectonics in accounting for the major physical features of the earth's surface. It is superbly illustrated and still highly regarded today.

———. *The Weather Machine.* London: British Broadcasting Corporation, 1974. 143 pages. ISBN: 0563126469.

A very readable account of meteorological phenomena and events by an expert at putting science into everyday language.

Carson, Rachel. *Silent Spring.* Boston: Houghton Mifflin; Cambridge, MA: Riverside Press, 1962. 368 pages.

The one book that likely more than any other kick-started today's environmental movement. Rachel Carson, a marine biologist and environmentalist, drafted this passionate exposé about the impact of pesticides on the environment when she was already seriously ill with cancer. The book caused a furore, and Carson spent her last months under bitter attack by industry representatives. She helped change the outlook of a generation. Beautifully written and cogently argued, the work remains fresh almost four decades later.

Carter, Bill. *Coastal Environments: An Introduction to the Physical, Ecological, and Cultural Systems of Coastlines.* London, New York: Academic Press, 1988. 617 pages. ISBN: 0121318552

An academic overview.

Caufield, Catherine. *In the Rainforest.* New York: Knopf, 1984. 304 pages. ISBN: 03945270111.

Repeated opinion surveys show that tropical deforestation is the single environmental issue that causes the most concern around the world. This book, written by an American journalist, is the account of a personal investigation into the causes and effects of tropical forest loss, covering Africa, Asia, and Latin America.

Chisholm, Michael. *Human Geography: Evolution or Revolution?* Harmondsworth, Baltimore: Penguin, 1975. 207 pages. ISBN: 0140218831.

In the 1960s and 1970s there were a number of important developments and innovations in the nature of geography and this book is Chisholm's attempt "to convey an account of the direction and purpose of recent changes in human geography as conceived by someone fairly close to the scene." Chisholm's book is a fascinating account that traces the history of the subject to the mid-1970s and attempts to look ahead into what was then regarded as a very uncertain future.

Cowen, Richard. *History of Life.* Boston: Blackwell Scientific Publications, 1990. 470 pages. ISBN: 0865420726.

A delightfully good-humored account, complete with arcane limericks and other incidental entertainment, which highlight the science of evolutionary biology without distracting from it. Good on invertebrates. A radical expansion on the first edition.

Crosby, Alfred W. *Ecological Imperialism: The Biological Expansion of Europe, 900–1900.* Cambridge, New York: Cambridge University Press, 1986. 368 pages. ISBN: 0521320097.

Looks at the biological expansion of Europe over 1,000 years, with particular emphasis on the colonial period, and argues that displacement of native peoples by Europeans in most temperate regions was as much driven by ecology as military conquest. Interesting for its perspectives linking ecology with historical movements.

Dawson, Alastair. *Ice Age Earth: Late Quaternary Geology and Climate.* London, New York: Routledge, 1992. 293 pages. ISBN: 0415015669.

A detailed account of the meteorological and geological evidence for the changes in the climate during the Ice Age. What we know about the current changes in the world's climate is in great measure reliant on what we know about past climatic change. The Ice Age changes in climate were not man-made. Today, in addition to man's interference with the climate, there are still many natural causes of global warming and cooling.

Decker, R. W., and B. B. Decker. *Mountains of Fire: The Nature of Volcanoes.* Cambridge, New York: Cambridge University Press, 1991. 198 pages. ISBN: 052132176X.

A well-illustrated introduction to volcanoes, their rock types, and their effects.

Dixon, Dougal. *The Practical Geologist.* New York: Simon and Schuster, 1992. 160 pages. ISBN: 0671746987.

A useful coverage of the various aspects of geology and the techniques used to study them.

Dunning, F. W. *The Story of the Earth* (1981). London: Natural History Museum, 1981. 36 pages. ISBN: 011310023X.

Gives an overview of the concept of plate tectonics and shows how volcanoes and earthquakes fit into the grand scheme.

Durning, Alan Thein. *How Much Is Enough? The Consumer Society and the Future of the Earth.* New York: Norton, 1992. 200 pages. ISBN: 039303383X.

Addresses the of how the rapidly increasing levels of consumption—of raw materials, energy, and other resources—by the rich fifth of the population can be reconciled with the needs of the environment and of the poorer human majority.

Emery, Dominic, and Keith Meyers, eds. *Sequence Stratigraphy.* Oxford, Cambridge, MA: Blackwell Scientific Publications, 1996. 297 pages. ISBN: 0632037067.

Excellent textbook derived from an in-house course run by international oil company British Petroleum.

Fortey, Richard. *Fossils: The Key to the Past.* Cambridge, MA: Harvard University Press, 1991. 187 pages. ISBN: 06743111353.

The author brings fossils to life and applies this knowledge to geology and biology. An accessible overview of all the branches of the subject, introducing all the main fossil organisms at the same time.

Fowler, Cary, and Pat Mooney. *The Threatened Gene: Food, Politics and the Loss of Genetic Diversity.* Tucson: University of Arizona Press, 1990. 278 pages. ISBN: 0816511543.

Biodiversity loss doesn't only affect natural systems; during the current century we have lost hundreds of traditional crop strains and livestock breeds. This meticulously researched and readable book looks at what has happened, how genetic diversity has been lost in food production, and why it matters.

Gage, J. D., and A. Tyler. *Deep-Sea Biology — A Natural History of Organisms at the Deep-Sea Floor.* Cambridge; New York: Cambridge University Press, 1991. 504 pages. ISBN: 0521334314.

The deep sea is one of the largest ecosystems, covering some 50 percent of the planet. This scholarly work introduces the reader to how oceanographers study the animals and their ecology in this alien world, where the temperature is only just above freezing and the pressure reaches several metric tons per square centimeter.

Goudie, Andrew, and Rita Gardner. *Discovering Landscape in England and Wales.* London, Boston: G. Allen & Unwin, 1985. 177 pages. ISBN: 045510768.

This book aims to "discover and try to explain some of the most appealing features of the natural landscape." It is a splendid book for all those who want to know a bit more about the history and geology of well-known British sites such as Helvellyn, Lulworth Cove, and Cheddar Gorge.

Gould, Stephen Jay, ed. *The Book of Life.* New York: W.W. Norton, 1993. 256 pages. ISBN: 0393035573.

The thinking person's paleontological coffee table book (and why not? See Gould's panegyric on the significance of coffee tables and their books). First-class illustrations and chapters by an impressive cast list of other distinguished specialists with emphasis on vertebrates, including humans.

———. *Wonderful Life: The Burgess Shale and the Nature of History.* New York: W.W. Norton, 1989. 347 pages. ISBN: 0393027058.

Gould is paleontology's most virtuoso narrator and interpreter. Here he brings to life in epic style the famously mysterious, and miraculously preserved, animals of the Burgess Shale, and the personalities who have studied them, using them as the raw material for expounding his ideas about the significance of chance.

Groombridge, Brian, ed. *Global Biodiversity: The Status of the Earth's Living Resources.* London, New York: Chapman and Hall, 1992. 585 pages. ISBN: 0412472406.

A massive review of biodiversity around the world, looking at the status of key ecosystems, the human uses of plants and animals, rates of change in natural systems, and so on.

Hardy, Alister. *The Open Sea— Its Natural History, Part 1: The World of the Plankton.* London: Fontana, 1972.

This is a classic text, wonderfully written, which will serve as a good introduction to the open sea and the intriguing creatures that live in it. Slightly dated now but nevertheless a fascinating account.

Hardy, Ralph, Peter Wright, John Gribbin, and John Kington. *The Weather Book.* Boston: Little, Brown, 1982. 224 pages. ISBN: 0316346233.

A book to fascinate anyone with an interest in meteorology. Packed with color pictures and amazing facts.

Harlow, George E. *The Nature of Diamonds.* Cambridge, New York: Cambridge University Press, 1998. 278 pages. ISBN: 052162083X.

This book takes the reader to the frontiers of diamond exploration and exploitation, offering a glimpse into the economics of the diamond trade. Both an informative and very readable book.

Holmes, Arthur. *The Principles of Physical Geology.* New York: Ronald Press Co., 1965. 1,288 pages.

The standard university textbook that deals principally with landscape formation.

Kaharl, Victoria A. *Water Baby: The Story of Alvin.* New York: Oxford University Press, 1990. 356 pages. ISBN: 0195061918.

This is a fascinating account of the building and development of the research submersible *Alvin*. She is one of a handful of vehicles which takes scientists down into the depths to glimpse this alien realm. *Alvin* has been the platform from which researchers have made many startling discoveries. It is the oceanographic equivalent of the space shuttle.

Lear, Linda. *Rachel Carson: Witness for Nature.* New York: H. Holt, 1997. 634 pages. ISBN: 0805034277.

The first full biography, it describes both Carson's life and career, and also her influence on the environmental movement.

Lovelock, James. *Gaia: A New Look at Life on Earth.* Oxford: Oxford University Press, 1982. 176 pages. ISBN: 0192860305.

Proposes the still heretical theory that the planet itself is, biologically speaking, a self-regulating mechanism. Whether you agree with Lovelock or not, he presents a sober and well-argued case, and his book helped focus attention on the role that different components play in the ecology of the planet, thus paving the way for discussion on biodiversity.

Lovins, Amory. *Soft Energy Paths: Towards a Durable Peace.* San Francisco: Friends of the Earth International, 1977. 231 pages. ISBN: 0884106144.

At a time when solar and wind power were regarded as little more than utopian dreams, Lovins turned the argument around and boldly claimed that, using established technologies, renewable sources will supply all our energy in the future. Part research paper and part polemic, *Soft Energy Paths* provided much of the inspiration for the continuing development of wind farms, solar villages, and small-scale hydropower projects. The title became synonymous with post-oil energy sources.

MacArthur, Robert, and Edward O. Wilson. *The Theory of Island Biogeography.* Princeton, NJ: Princeton University Press, 1967. 203 pages.

This seminal text presents and explores the relationship between the number of species on an island and its size, distance from the mainland, and biogeographical history. The book has stimulated more research in conservation than any other work and remains controversial.

McNeely, Jeffrey A., et al. *Conserving the World's Biological Diversity.* Gland, Switzerland: International Union for Conservation of Nature and Natural Resources, 1990. 193 pages. ISBN: 0915825422.

The most comprehensive attempt so far to set an agenda for international conservation of biodiversity. Much of the thinking was later incorporated into proposals for the Convention on Biological Diversity.

Meadows, Donella L., Dennis L. Meadows, Jorgen Randers, and William W. Behrens III. *The Limits to Growth.* New York: Universe Books, 1975. 205 pages. ISBN: 0876631650.

An early and influential attempt to map the likely impacts of resource depletion using computer modeling. Coming immediately after the 1973 oil crisis, the book helped create new attitudes towards nonrenewable resources. The predictions have proved to be overly pessimistic in terms of time scale, although the central argument remains valid, and the team produced a new analysis, *Beyond the Limits,* almost 20 years later.

Moore, George W. *Speleology: Caves and the Cave Environment.* St. Louis: Cave Books, 1997. 176 pages. ISBN: 0939748460.

A general and wide-ranging look at cave science that will draw in the general reader.

Myers, Norman. *The Gaia Atlas of Planet Management.* New York: Anchor Books, 1993. 272 pages. ISBN: 0385426267.

There are many atlases and reference books on the market today but there are few that are as informative and lavishly illustrated as this. This thought-provoking work provides many excellent thematic maps and color photographs. It is divided into several sections including the land, the oceans, the elements, evolution, and civilization. Strongly recommended for all those with an interest and concern for the issues affecting the future of our planet.

———. *The Sinking Ark.* Oxford, New York: Pergamon Press, 1979. 307 pages.

Argues that we face a massive loss of species, particularly through tropical deforestation, and that conservation efforts are manifestly failing to counter this trend. In this book Myers also puts forth the practical case for preserving species, in terms of their use as genetic resources for foodstuffs, medicines, and so on, a theme which he has continued to developed in a number of further titles, including *The Primary Source.*

Norse, Elliott A., ed. *Global Marine Biological Diversity.* Washington DC: Island Press, 1993. 383 pages. ISBN: 1559632550.

It is often forgotten that the sea represents one of the most diverse systems on the planet. This book explains the problems facing conserving and sustaining biodiversity in the marine environment. It provides a readable account of the current debate about biodiversity and how and why we should try to sustain it.

Noss, Reed F., and Allen Y. Cooperrider. *Saving Nature's Legacy.* Washington DC: Island Press, 1994. 416 pages. ISBN: 155963247X.

Guidelines on practical biodiversity conservation at a landscape level.

Pellant, Chris. *Rocks, Minerals and Fossils of the World.* Boston: Little, Brown, 1990. 175 pages. ISBN: 0316697974.

It is normally not a good idea to base mineral identification on photographs—the diagnostic properties do not show. However, the photographs in this book are particularly good.

Ponting, Clive. *The Green History of the World.* New York: St. Martin's Press, 1992. 432 pages. ISBN: 0312069871.

A history of the world from an environmental standpoint, packed with information but also presenting a sustained argument about the links between ecology and human activities, and linking in the influence of religious thought and political philosophy. An excellent one-volume introduction.

Raup, David. *The Nemesis Affair. A Story of the Death of Dinosaurs and Ways of Science.* New York: Norton, 1986. 220 pages. ISBN: 03930234227.

Treats extinction generally, and discusses controversial collisions between extraterrestrial bodies and earth, between organisms and their physical world, and between scientists and mass culture.

Reading, Harold. *Sedimentary Environments: Processes, Facies and Stratigraphy*. Cambridge, MA: Blackwell Scientific Publications, 1996. ISBN: 0632036273.

Possibly the best guide to understanding sedimentary rocks.

Rudwick, Martin. *Georges Cuvier, Fossil Bones, and Geological Catastrophes.* Chicago: University of Chicago Press, 1997. 301 pages. ISBN: 0226731065.

A guide to his most important works.

———. *The Meaning of Fossils.* New York: American Elsevier, 1972. 287 pages. ISBN: 0444195769.

Widely regarded as the standard scholarly reference to the historical scientific issues and intellectual questions posed by paleontology.

Sabins, Floyd. *Remote Sensing: Principles and Interpretation.* New York: W. H. Freeman and Co., 1997. 494 pages. ISBN: 0716724421.

Remote sensing entails the gathering of scientific data from a distance; in earth science, this is generally done by probes and satellites. A comprehensive textbook that is also entertaining.

Shiva, Vandana. *Biodiversity: Social and Ecological Perspectives.* London, Atlantic Highlands, NJ: Zed Books. 123 pages. ISBN: 185649053X.

Proposes that biodiversity conservation has to be tackled by first addressing problems of intensive farming and North-South relations; one of a series of books by a noted Indian scientist and feminist.

Smith, David G., ed. *The Cambridge Encyclopedia of Earth Sciences.* New York: Crown Publisher/ Cambridge University Press, 1981. 496 pages. ISBN: 0517543702.

Rather academic treatment of all the physical earth sciences (of which geology is a part), but well illustrated.

Smith, James. *Introduction to Geodesy: The History and Concepts of Modern Geodesy.* New York: Wiley, 1996. ISBN: 047116660X.

Geodesy concerns methods of surveying the earth for making maps and correlating geological, gravitational, and magnetic measurements. This work offers a history of geodesy that is accessible to the general reader.

Stanley, S. M. *Earth and Life through Time.* New York: W. H. Freeman, 1986. 690 pages. ISBN: 0716716771.

Earth and life and the meaning of nearly everything. This is the American college solution to everything you wanted to know about fossils, evolution, and geology. Don't worry about the self-assessment questions unless you are a student—this book is here on its encyclopedic merit.

Swinnerton, H. H. *Fossils.* London: Collins, 1960. 274 pages.

A notable contribution to the classic *New Naturalist* series, so inevitably a very homely British bias. No matter—the range of fossil-bearing rocks in Britain is a good enough sample of the fossil record, which Swinnerton sets out with charm, enjoyment, and enthusiasm.

Tassy, Pascal. *The Message of Fossils.* New York: McGraw-Hill, 1993. 163 pages. ISBN: 0070629471.

An easygoing essay focused on the interface between paleontology and evolution, which includes its more recent controversies.

Thackray, John. *The Age of the Earth.* London: Her Majesty's Stationary Office for the Institute of Geological Sciences, 1980. 34 pages. ISBN: 0118840770.

An excellent rundown on the techniques used to work out geological time.

Wells, Sue, and Nick Hanna. *The Greenpeace Book of Coral Reefs.* New York: Sterling Publishing Co., 1992. 160 pages. ISBN: 0806987952.

The main thrust of this book is explaining the coral ecosystem and the complex cycles of life that are increasingly under threat by human activities.

Wendt, Herbert. *Before the Deluge.* Garden City, NY: Doubleday, 1970. 419 pages.

The history of paleontology, its issues and its personalities, rich in narrative and with a balanced international perspective.

Wilson, Edward O. *The Diversity of Life.* Cambridge, MA: Belknap Press of Harvard University Press, 1992. 424 pages. ISBN: 0674212983.

An outstanding overview of the biodiversity crisis, as well as basic concepts such as evolutionary change, extinction, and speciation, written for the general public. The all-encompassing range, the compelling case for conservation, and the delightful natural history are virtues enough to recommend this book.

———, ed. *Biodiversity* Washington DC: National Academy Press, 1988. 521 pages. ISBN: 0309037395.

The first book to crystallize the debate about biodiversity, which also brought the word itself to public attention. A fascinating array of articles, papers, and even poems about biological diversity around the world. Still the best single introduction to the subject. See also *The Diversity of Life* by Wilson himself for a readable introduction to the subject.

6 World Wide Web Sites

Arctic Circle
http://articcircle.uconn.edu/

Well-written site with information about all aspects of life in the Arctic. There are sections on history, natural resources, the rights of indigenous peoples, and issues of environmental concern.

Avalanche!
http://www.pbs.org/wgbh/nova/avalanche/

Companion to the U.S. Public Broadcasting Service (PBS) television program *Nova,* this page follows an intrepid documentary team as they set out to film an avalanche. It provides information on the causes of avalanches, as well as details on how the film crew avoided getting swept away by them. There are six video clips of actual avalanches in progress and a number of still photos. You can download a transcript of the television program.

Avalanche Awareness
http://www-nsidc.colorado.edu/NSIDC/EDUCATION/AVALANCHE/

Description of avalanches, what causes them, and how to minimize dangers if caught in one. There is advice on how to determine the stability of a snowpack, what to do if caught out, and how to locate people trapped under snow. Nobody skiing off a downhill ski trail should set off without reading this.

Badlands National Park
http://www.nps.gov/badl/exp/home.htm

Impressive U.S. National Park Service guide to the Badlands National Park. There is comprehensive information on the geology, flora and fauna, and history of the Badlands. Hikers are provided with detailed guidance on routes and there is a listing of the park's educational activities.

Before and After the Great Earthquake and Fire: Early Films of San Francisco, 1897–1916
http://lcweb2.loc.gov/ammem/papr/sfhome.html

Collection of 26 early films depicting San Francisco before and after the 1906 disaster, including a 1915 travelogue that shows scenes of the rebuilt city and the Panama Pacific Exposition, and a 1916 propaganda film. Should you not wish to download the entire film, each title contains sample still frames. There is also background information about the earthquake and fire, and a section of selected bibliography.

Big Empty
http://www.blm.gov/education/great_basin/great_basin.html

Sponsored by the U.S. Bureau of Land Management, this site explores the Great Basin of the western United States and its desert ecosystem of plants, animals, and minerals. This site includes information on the scarcity of water, modern and environmental challenges, mining, grazing, wild horses and burros, and a look at methods to preserve and rehabilitate the ecosystem. Maps and photographs complement the text.

Bodleian Library Map Room—The Map Case
http://www.rsl.ox.ac.uk/nnj/mapcase.htm

Broad selection of images from the historical map collection of the Bodleian Library at Oxford University. Visitors can choose among rare maps of Oxfordshire, London, areas of Britain, New England, Canada, and more. The maps can be viewed by thumbnail and then selected in their full GIF or JPEG versions.

British Geological Survey Global Seismology and Geomagnetism Group Earthquake Page
http://www.gsrg.nmh.ac.uk/gsrg.html

Fascinating maps showing the location and relative magnitude of recent U.K. earthquake activity make up just a small part of this website.

Also included are historical and archival information, as well as descriptions of felt effects and hazards. Watch out if you are in Wolverhampton!

Cambrian Period: Life Goes for a Spin
http://www.sciam.com/explorations/082597cambrian/powell.html

Part of a larger site maintained by *Scientific American,* this page reports on the research of Joseph Kirschvink of the California Institute of Technology, which suggests that the so-called Cambrian Explosion resulted from a sudden shifting of the earth's crust. The text includes hypertext links to further information, and there is also a list of related links, including one to figures, diagrams, and information from Kirschvink's research paper.

Cavediving
http://www.cavediving.com/home.htm

U.S. website run by a cave-diving instructor. The site provides all sorts of information for those who wish to swim in cold, dark, underwater caverns. Information on how and where to cave dive is provided along with news, articles, links, and details on where to learn how to cave dive.

Clouds and Precipitation
http://ww2010.atmos.uiuc.edu/(Gh)/guides/mtr/cld/home.rxml

Illustrated guide to how clouds form and to identifying the types of clouds. The site contains plenty of images and a glossary of key terms in addition to further explanations of the various types of precipitation.

Clouds from Space
http://www.hawastsoc.org/solar/eng/cloud1.htm

This site offers a unique look at clouds, with photographs of various cloud types taken from space, including thunderstorms over Brazil, jet-stream cirrus clouds, and a description of how clouds form.

Coriolis Effect
http://www.physics.ohio-state.edu/~dvandom/Edu/coriolis.html

Subtitled "a (fairly) simple explanation," this site contains a description of the principles involved, and is aimed at nonphysicists.

Cracking the Ice Age

http://www.pbs.org/wgbh/nova/ice/

Companion to the U.S. Public Broadcasting Service (PBS) television program *Nova,* this page provides information about glaciation, the natural changes in climate over the past 60 million years, the greenhouse effect, global warming, and continental movement. There is also a list of related links.

Daily Planet

http://www.atmos.uiuc.edu/

Masses of data and other information on climate around the planet. Among the site's more attractive features is the use of multimedia instructional modules, customized weather maps, and real-time weather data.

Dan's Wild Wild Weather Page

http://www.whnt19.com/kidwx/index.html

Introduction to the weather for kids. It has pages dealing with everything from climate to lightning, from satellite forecasting to precipitation; all explained in a lively style with plenty of pictures.

Decade of Notable Californian Earthquakes

http://www.earthwaves.com/shorose/califneq.html

Details of recent Californian earthquakes with maps, photographs, video clips, and animation. The site title is misleading as there is detailed information on earthquakes from 1980 to 1995, with official reports to back up the rendering of the seismic shocks in animations and video clips.

Double Whammy

http://www.sciam.com/explorations/1998/011998asteroid/

Part of a larger site maintained by *Scientific American,* this page explores the catastrophic impact that an asteroid's crashing into the sea would have on civilization and the environment. Animated simulations show the effects of impact and the effect that a tsunami would have on the eastern seaboard of the United States if an asteroid struck the Atlantic Ocean. Learn about the tsunami that struck Prince William Sound, Alaska, in 1964 after an underwater earthquake. The text includes hypertext links to further information and a list of related links.

Earth and Moon Viewer
http://www.fourmilab.ch/earthview/vplanet.html

View a map of the earth showing the day and night regions at this moment, or view the earth from the sun, the moon, or any number of other locations. Alternatively, take a look at the moon from the earth or the sun, or from above various formations on the lunar surface.

Earth Introduction
http://www.hawastsoc.org/solar/eng/earth.htm

Everything you ever wanted to know about planet Earth can be found at this site, which contains a table of statistics, photographs taken from space, radar-generated images of the planet, animations of the earth rotating, and more.

Earth's Seasons: Equinoxes, Solstices, Perihelion, and Aphelion
http://aa.usno.navy.mil/AA/data/docs/EarthSeasons.html

Part of a larger site on astronomical data maintained by the U.S. Naval Observatory, this site gives the dates and hours (in Universal Time) of the changing of the seasons from 1992 through 2005. It also includes sections of frequently asked questions (FAQs) and research information.

Edwards Aquifer Home Page
http://www.txdirect.net/users/eckhardt/

Guide to the Edwards Aquifer (a rock formation containing water) in Texas—one of the most prolific artesian aquifers in the world.

El Niño Theme Page
http://www.pmel.noaa.gov/toga-tao/el-nino/home.html

Wealth of scientific information about El Niño, a "disruption of the ocean-atmosphere system in the tropical Pacific," with animated views of the monthly sea changes brought about by it, El Niño–related climate predictions, and forecasts from meteorological centers around the world. It also offers an illuminating FAQ section with basic and more advanced questions as well as an interesting historical overview of the phenomenon, starting from 1550.

EqIIS—Earthquake Image Information System
http://www.eerc.berkeley.edu/cgi-bin/eqiis_form?eq=4570&count=1

Fully searchable library of almost 8,000 images from more than 80 earthquakes. It is possible to search by earthquake, structure, photographer, and keyword.

Flood!
http://www.pbs.org/wgbh/nova/flood/

Companion to the U.S. Public Broadcasting Service (PBS) television program *Nova,* this page concerns many aspects of flooding. It takes a historical look at floods and examines the measures that engineers have taken to combat them. Three major rivers are discussed: the Yellow, the Nile, and the Mississippi. In addition to learning about the negative effects of floods, visitors can also find out about the benefits that floods bestow on farmland. There are many images dispersed throughout the site, plus an audio clip of a flood in progress.

GeoClio
http://geoclio.st.usm.edu/

GeoClio is a webserver for the history of geology and the geosciences, so named for the muse of history, Clio, in Greek mythology. Established in 1995 by a grant from the National Science Foundation, the website provides historical information on the evolution of the geosciences and includes information on literature, meetings, discussion groups, and the Friends of GeoClio. The server is operated and maintained by the University of Southern Mississippi.

Geologylink
http://www.geologylink.com/

Comprehensive information on geology that features a daily update on current geologic events, virtual classroom tours, and virtual field trips to locations around the world. Visitors will also find an in-depth look at a featured event, geologic news and reports, an image gallery, glossary, maps, and an area for asking geology professors their most perplexing questions, plus a list of references and links.

Glaciers
http://www-nsidc.colorado.edu/NSIDC/EDUCATION/GLACIERS/

Comprehensive information about glaciers from the U.S. National Snow and Ice Data Center. There are explanations of why glaciers form, different kinds of glaciers, and what they may tell us about climate change. There are a number of interesting facts and a bibliography about the compacted tongues of ice that cover 10 percent of the land surface of our planet.

Hurricane & Tropical Storm Tracking
http://hurricane.terrapin.com/

Follow the current paths of Pacific and mid-Atlantic hurricanes and tropical storms at this site. Java animations of storms in previous years can also be viewed, and data sets for these storms may be downloaded. Current satellite weather maps can also be accessed, but only for the United States and surrounding region.

Hydrology Primer
http://wwwdmorll.er.usgs.gov/~bjsmith/outreach/hydrology.primer.html

Information from the U.S. Geological Survey about all aspects of hydrology. The clickable chapters include facts about surface water and ground water, the work of hydrologists, and careers in hydrology. For answers to further questions click on "ask a hydrologist," which provides links to other U.S. national and regional sources.

Katmai National Park and Preserve
http://www.alaskanet.com/Tourism/Parks/Katmai/

Guide to the National Park, which contains the famous Valley of Ten Thousand Smokes, where the twentieth century's most dramatic volcanic episode took place.

Late Pleistocene Extinctions
http://www.museum.state.il.us/exhibits/larson/LP_extinction.html

Exploration of possible causes of the Late Pleistocene extinction of most large mammals in North America. Different theories are discussed. There is also information on the prehistoric inhabitants of North America.

Met. Office Home Page
http://www.meto.govt.uk/

Authoritative account of global warming issues such as the ozone problem, El Niño (and the lesser-known La Niña that follows the former), the tropical cyclones, and forecasting methods. Scientific explanations alternate with images and film clips in an educational site that especially targets teachers and their students.

MTU Volcanoes Page
http://www.geo.mtu.edu/volcanoes/

Provided by Michigan Technological University, this site includes a world map of volcanic activity with information on recent eruptions, the latest research in remote sensing of volcanoes, and many spectacular photographs.

Multimedia History of Glacier Bay, Alaska
http://sdcd.gsfc.nasa.gov/GLACIER.BAY/glacierbay.html

Virtual tour featuring scenic flights over three-dimensional glaciers. The multimedia experience combines video footage, computer and satellite images, photos, text, and maps, and includes information on glacial formation. View hand-drawn maps of glaciers that date back to 1794 and watch video clips of massive ice fronts splitting and splashing into the sea.

Museum of Paleontology
http://www.ucmp.berkeley.edu/exhibit/exhibits.html

Large amount of detailed information on the subject in a carefully structured and carefully cross-referenced site. Visitors can explore paleontology through the three areas of phylogeny, geology, and evolution.

Ocean Planet
http://seawifs.gsfc.nasa.gov/ocean_planet.html

Oceans and the environmental issues affecting their health, based on the Smithsonian Institution's traveling exhibition of the same name. Use the exhibition floor plan to navigate your way around the different "rooms"—with themes ranging from Ocean Science and Immersion to Heroes and Sea People.

Ocean Satellite Image Comparison
http://www.csc.noaa.gov/crs/real_time/javaprod/satcompare.html

Maintained by the U.S. National Oceanic Atmospheric Administration, this site is a Java applet that compares satellite images and data about the ocean surface temperatures and turbidity of numerous coastal areas around the United States. The site includes an Image Panner for navigating larger images.

Plate Tectonics
http://www.seismo.unr.edu/ftp/pub/louie/class/100/plate-tectonics.html

Well-illustrated site on this geological phenomenon. In addition to the plentiful illustrations, this site has a good, clear manner of explaining the way the plates of the earth's crust interact to produce seismic activity.

Questions and Answers about Snow
http://www-nsidc.colorado.edu/NSIDC/EDUCATION/SNOW/snow_FAQ.html

Comprehensive information about snow from the U.S. National Snow and Ice Data Center. Among the interesting subjects discussed are why snow is white, why snowflakes can be up to two inches across, what makes some snow fluffy, why sound travels farther across snowy ground, and why snow is a good insulator.

San Andreas Fault and Bay Area
http://sepwww.stanford.edu/oldsep/joe/fault_images/BayAreaSanAndreasFault.html

Detailed tour of the San Andreas Fault and the San Francisco Bay Area, with information on the origination of the fault. The site is supported by a full range of area maps.

Strange Science
http://www.turnpike.net/~mscott/index.htm

Subtitled "The Rocky Road to Modern Paleontology and Biology," this site examines some of the early, medieval discoveries that led to the growth of interest in modern-day science. The site is illustrated with images that clearly show how people's perception of the world differed, and how people made up for gaps in their knowledge with a little imagination!

Tsunami!
http://www.geophys.washington.edu/tsunami/intro.html

Description of many aspects of tsunamis. Included are details on how a tsunami is generated and how it propagates, how they have affected humans, and how people in coastal areas are warned about them. The site also discusses if and how people can protect themselves from tsunamis and provides "near real-time" tsunami information bulletins.

Virtual Cave
http://www.vol.it/MIRROR2/EN/CAVE/virtcave.html

Browse the mineral wonders unique to the cave environment—from bell canopies and bottlebrushes to splattermites and stalactites.

VolcanoWorld
http://volcano.und.edu

Comprehensive, fully searchable site on volcanoes that includes details of the most recent eruptions, the volcanoes that are currently active, a glossary, images and video clips, and even a list of extraterrestrial volcanoes. If visitors can't find out what they want to know, they are even encouraged to "ask a volcanologist."

Welcome to Coral Forest
http://www.blacktop.com/coralforest/

Site dedicated to explaining the importance of coral reefs for the survival of the planet. It is an impassioned plea on behalf of the world's endangered coral reefs and includes a full description of their biodiversity, maps of where coral reefs are to be found (no less than 109 countries), and many photos.

Windows to the Universe
http://www.windows.umich.edu/

NASA-funded project providing detailed information about both earth and space sciences. The information is divided into three levels: beginner, intermediate, and advanced, and covers such topics as the planets in our solar system, the earth and its atmosphere, and also links to sites with information on earthquakes and volcanoes, precipitation and sea ice, and storms and hurricanes.

Woods Hole Oceanographic Institution Home Page
http://www.whoi.edu/index.html

Site run by a Massachusetts-based oceanographic institute. As well as containing details of its research programs and an overview of the organization, the institute provides a gallery of marine pictures and videos, and contacts for its education programs.

World Meteorological Organization
http://www.wmo.ch/

Internet voice of the World Meteorological Organisation (WMO), a United Nations division coordinating global scientific activity related to climate and weather. The site offers ample material on the long-term objectives and immediate policies of the organization. It also disseminates important information on WMO's databases, training programs, and satellite activities, as well as its projects related to the protection of the environment.

World of Amber
http://www.emporia.edu/S/www/earthsci/amber/amber.htm

Everything you need to know about amber is here. There is information on its physical properties, uses, and geological and geographical occurrences, plus fossils in amber, recovery methods, amber myths, museums, and a quiz.

Worldwide Earthquake Locator
http://www.geo.ed.ac.uk/quakes/quakes.html

Edinburgh University, Scotland, runs this site that allows visitors to search for the world's latest earthquakes. The locator works on a global map on which perspective can be zoomed in or out. There are normally around five or six earthquakes a day—you'll find it surprising how few make the news. The site also has some general information on earthquakes.

Innovations in Earth Sciences

Part II

Dictionary of Terms and Concepts

ablation

The loss of snow and ice from a glacier by melting and evaporation. It is the opposite of accumulation. Ablation is most significant near the snout, or foot, of a glacier, because temperatures tend to be higher at lower altitudes. The rate of ablation also varies according to the time of year, being greatest during the summer. If total ablation exceeds total accumulation for a particular glacier, the glacier will retreat. If total accumulation exceeds ablation, the glacier will advance.

abrasion

The effect of corrasion, a type of erosion in which rock fragments scrape and grind away a surface. The rock fragments may be carried by rivers, wind, ice, or the sea. Striations, or grooves, on rock surfaces are common abrasions, caused by the scratching of rock debris embedded in glacier ice.

abrasive

From the Latin for *scraping away.* A substance used for cutting and polishing or for removing small amounts of the surface of hard materials. There are two types of abrasives: natural and artificial. Their hardness is measured using the Mohs scale. Natural abrasives include quartz, sandstone, pumice, diamond, emery, and corundum. Artificial abrasives include rouge, whiting, and carborundum.

abyssal plain

A broad expanse of seafloor lying 3–6 km/2–4 mi below sea level. Abyssal plains are found in all the major oceans, and they extend from bordering continental rises to midoceanic ridges. Underlain by outward-spreading new oceanic crust extruded from ridges, abyssal plains are covered in deep-sea sediments derived from continental slopes and floating microscopic marine organisms. The plains are interrupted by chains of volcanic islands and seamounts, where plates have ridden over hot spots in the mantle, and by additional seamounts, which were originally formed in oceanic ridge areas and transferred to the deep as the oceanic crust subsided. Otherwise, the abyssal plains are very flat, with a slope of less than 1:1000.

abyssal zone

A dark ocean region 2,000–6,000 m/6,500–19,500 ft deep; temperature 4°C/39°F. Three-quarters of the area of the deep ocean floor lies in the abyssal zone, which is too far from the surface for photosynthesis to take place. Some fish and crustaceans living there are blind or have their own light sources. The region above is the bathyal zone; the region below, the hadal zone.

accumulation

The addition of snow and ice to a glacier. It is the opposite of ablation. Snow is added through snowfall and avalanches, and is gradually compressed to form ice. Although accumulation occurs at all parts of a glacier, it is most significant at higher altitudes near the glacier's start, where temperatures are lower.

acid rain

Acidic precipitation thought to be caused principally by the release into the atmosphere of sulfur dioxide (SO_2) and oxides of nitrogen, which dissolve in pure rainwater and make it acidic. Sulfur dioxide is formed by the burning of fossil fuels, such as coal, that contain high quantities of sulfur; nitrogen oxides are contributed from various industrial activities and from car exhaust fumes.

Acidity is measured on the pH scale, in which the value of 0 represents liquids and solids that are completely acidic and 14 represents those that are highly alkaline. Distilled water is neutral and has a pH of 7. Normal rain has a value of 5.6. It is slightly acidic due to the presence of carbonic acid formed by the mixture of CO_2 and rainwater. Acid rain has values of 5.6 or less on the pH scale.

Acid deposition occurs not only as wet precipitation (mist, snow, or rain), but also comes out of the atmosphere as dry particles (dry deposition) or is absorbed directly by lakes, plants, and masonry as gases. Acidic gases can travel over 500 km/310 mi a day, so acid rain can be considered an example of transboundary pollution. Acid rain is linked with damage to and the death of forests and lake organisms in Scandinavia, Europe, and eastern North America. It also results in damage to buildings and statues. U.S. and European power stations that burn fossil fuels release about 8 g/0.3 oz of sulfur dioxide and 3 g/0.1 oz of nitrogen oxides per kilowatt-hour. According to the U.K. Department of the Environment figures, emissions of sulfur dioxide from power stations would have to be decreased by 81 percent in order to arrest damage. To reduce the amount of acid rain falling on North America, U.S. power companies cut their sulfur dioxide output by 5.6 million tons in 1995.

The main effect of acid rain is to damage the chemical balance of soil, causing leaching of important minerals including magnesium and aluminum. Plants living in such soils, particularly conifers, suffer from mineral loss and become more prone to infection. The minerals from the soil pass into lakes and rivers, disturbing aquatic life, for instance, by damaging the gills of young fish and killing plant life. Lakes affected by acid rain are virtually clear due to the absence of green plankton. Lakes and rivers suffer more direct damage as well because they become acidified by rainfall draining directly from their catchment.

Owing to reductions in sulfur emissions, the amount of sulfur deposited per hectare of farmland has fallen from around 50 kg/110 lb in 1979 to 10 kg/22 lb in 1995. According to German research in 1995, although acid rain is harmful to crops, it also provides essential sulfur. There has been an increased incidence of sulfur deficiency diseases in plants from 1990 to 1995.

acid rock

Igneous rock that contains more than 60 percent by weight silicon dioxide, SiO_2, such as a granite or rhyolite. Along with the terms basic rock and ultrabasic rock, it is part of an outdated classification system based on the erroneous belief that silicon in rocks is in the form of silicic acid. Geologists today are more likely to use the descriptive term *felsic rock* or report the amount of SiO_2 in weight percent.

aclinic line

The magnetic equator, an imaginary line near the equator, where a compass needle balances horizontally, the attraction of the north and south magnetic poles being equal.

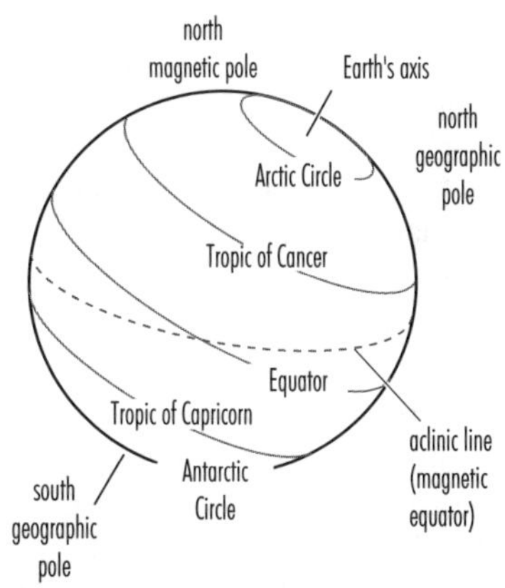

advection fog

Fog formed by warm moist air meeting a cold ocean current or flowing over a cold land surface. It is common in coastal areas where warm air from the sea meets the colder land surface, particularly at night or during the winter.

agate

Cryptocrystalline (with crystals too small to be seen with an optical microscope) silica, SiO_2, composed of cloudy and banded chalcedony, sometimes mixed with opal, that forms in rock cavities. Agate stones, being hard, are also used to burnish and polish gold applied to glass and ceramics and as clean vessels for grinding rock samples in laboratories.

air mass

A large body of air with particular characteristics of temperature and humidity. An air mass forms when air rests over an area long enough to pick up the conditions of that area. When an air mass moves to another area it affects the weather of that area, but its own characteristics become modified in the process. For example, an air mass formed over the Sahara will be hot and dry, becoming cooler as it moves northward. Air masses that meet form fronts.

There are four types of air masses: Tropical continental (Tc) air masses form over warm land and Tropical maritime (Tm) masses form over warm seas. Air masses that form over cold land are called Polar continental (Pc) and those forming over cold seas are called Polar or Arctic maritime (Pm or Am).

airglow

Faint and variable light in the earth's atmosphere produced by chemical reactions (the recombination of ionized particles) in the ionosphere.

alabaster

A naturally occurring fine-grained white or light-colored translucent form of gypsum, often streaked or mottled. A soft material, it is easily carved, but seldom used for outdoor sculpture.

alexandrite

A rare gemstone variety of the mineral chrysoberyl (beryllium aluminum oxide, $BeAl_2O_4$), which is green in daylight but appears red in artificial light.

alluvial deposit

A layer of broken rocky matter, or sediment, formed from material that has been carried in suspension by a river or stream and dropped as the velocity of the current decreases. River plains and deltas are made entirely of alluvial deposits, but smaller pockets can be found in the beds of upland torrents.

Alluvial deposits can consist of a whole range of particle sizes, from boulders down through cobbles, pebbles, gravel, sand, silt, and clay. The raw materials are the rocks and soils of upland areas that are loosened by erosion and washed away by mountain streams. Much of the world's richest farmland lies on alluvial deposits. These deposits can also provide an economic source of minerals. River currents produce a sorting action, with particles of heavy material deposited first while lighter materials are washed downstream. Hence, heavy minerals such as gold and tin, present in the original rocks in small amounts, can be concentrated and deposited on stream beds in commercial quantities. Such deposits are called placer ores.

alluvial fan

A roughly triangular sedimentary formation found at the base of slopes. An alluvial fan results when a sediment-laden stream or river rapidly deposits its load of gravel and silt as its speed is reduced on entering a plain. The surface of such a fan slopes outward in a wide arc from an apex at the mouth of the steep valley. A small stream carrying a load of coarse particles builds a shorter, steeper fan than a large stream carrying a load of fine particles. Over time, the fan tends to become destroyed piecemeal by the continuing headward and downward erosion leveling the slope.

alluvium

Fine silty material deposited by a river. It is deposited along the river channel where the water's velocity is too low to transport the river's load—for example, on the inside bend of a meander. A flood plain is composed of alluvium periodically deposited by floodwater. Sometimes alluvial deposits after flooding result in oxbow lakes. A deposit at the mouth of a river entering the sea is known as marine alluvium.

Flood plains are often fertile, but because they are liable to flooding, they may be of only limited agricultural value. Nevertheless, the major early civilizations developed in the flood plains of the Middle East where silts brought down by rivers such as the Nile and the Indus allowed a food surplus to be grown.

altimetry

A method of measuring changes in sea level using an altimeter attached to a satellite. The altimeter measures the distance between the satellite and water surface.

amethyst

A variety of quartz, SiO_2, colored violet by the presence of small quantities of impurities such as manganese or iron; used as a semiprecious stone. Amethysts are found chiefly in the Ural Mountains, India, the United States, Uruguay, and Brazil.

amphibole

Any one of a large group of rock-forming silicate minerals with an internal structure based on double chains of silicon and oxygen, and with a general formula $X_2Y_5Si_8O_{22}(OH)_2$; closely related to pyroxene. Amphiboles form orthorhombic, monoclinic, and triclinic crystals. Amphiboles occur in a wide range of igneous and metamorphic rocks. Common examples are hornblende (X = Ca, Y = Mg, Fe, Al) and tremolite (X = Ca, Y = Mg).

anabatic wind

A warm wind that blows uphill in steep-sided valleys in the early morning. As the sides of a valley warm up in the morning the air above is also warmed and rises up the valley to give a gentle breeze. By contrast, a katabatic wind is cool and blows down a valley at night.

anabranch

From the Greek *ana,* meaning *again.* A stream that branches from a main river, then reunites with it. For example, the Great Anabranch in New South Wales, Australia, leaves the Darling near Menindee, and joins the Murray below the Darling-Murray confluence.

andalusite

Aluminum silicate, Al_2SiO_5, a white to pinkish mineral crystallizing as square- or rhombus-based prisms. It is common in metamorphic rocks formed from clay sediments under low pressure conditions. Andalusite, kyanite, and sillimanite are all polymorphs of Al_2SiO_5.

andesite

A volcanic igneous rock, intermediate in silica content between rhyolite and basalt. It is characterized by a large quantity of feldspar minerals, giving it a light color. Andesite erupts from volcanoes at destructive plate margins (where one plate of the earth's surface moves beneath another; *see* **plate tectonics**), including the Andes, from which it gets its name.

Antarctic Circle

An imaginary line that encircles the South Pole at latitude 66° 32' S. The line encompasses the continent of Antarctica and the Antarctic Ocean. The region south of this line experiences at least one night in the southern summer during which the sun never sets, and at least one day in the southern winter during which the sun never rises.

anthracite

From the Greek *anthrax,* meaning *coal.* A hard, dense, shiny variety of coal, containing over 90 percent carbon and a low percentage of ash and impurities, which causes it to burn without flame, smoke, or smell. Because of its purity, anthracite gives off relatively little sulfur dioxide when burned. Anthracite gives intense heat, but is slow burning and slow to light; it is therefore unsuitable for use in open fires. Its characteristic composition is thought to be due to the action of bacteria in disintegrating the coal-forming material when it was laid down during the Carboniferous period. Among the chief sources of anthracite coal are Pennsylvania; southern Wales; the Donbas in Ukraine and Russia; and Shanxi province in China.

anticline

Rock layers or beds folded to form a convex arch (seldom preserved intact) in which older rocks comprise the core. Where relative ages of the rock layers, or stratigraphic ages, are not known, convex upward folded rocks are referred to as *antiforms.* The fold of an anticline may be undulating or steeply curved. A steplike bend in otherwise gently dipping or horizontal beds is a monocline. The opposite of an anticline is a syncline.

anticyclone

An area of high atmospheric pressure caused by descending air, which becomes warm and dry. Winds radiate from a calm center, taking a clockwise direction in the Northern Hemisphere and a counterclockwise direction in the Southern

Hemisphere. Anticyclones are characterized by clear weather and the absence of rain and violent winds. In summer they bring hot, sunny days and in winter they bring fine, frosty spells. Blocking anticyclones, which prevent the normal air circulation of an area, can cause summer droughts and severe winters.

antipodes

Greek for *opposite feet.* Places at opposite points on the globe.

apatite

A common calcium phosphate mineral, $Ca_5(PO_4)_3(F,OH,Cl)$. Apatite has a hexagonal structure and occurs widely in igneous rocks, such as pegmatite, and in contact metamorphic rocks, such as marbles. It is used in the manufacture of fertilizer and as a source of phosphorus. Carbonate hydroxylapatite, $Ca_5(PO_4,CO_3)_3(OH)_2$, is the chief constituent of tooth enamel and, together with other related phosphate minerals, is the inorganic constituent of bone. Apatite ranks 5 on the Mohs scale of hardness.

Appleton layer

Also *F layer.* A band containing ionized gases in the earth's upper atmosphere, at a height of 150–1,000 km/94–625 mi, above the E layer (formerly the Kennelly-Heaviside layer). It acts as a dependable reflector of radio signals as it is not affected by atmospheric conditions, although its ionic composition varies with the sunspot cycle. The Appleton layer has the highest concentration of free electrons and ions of the atmospheric layers. It is named for the English physicist Edward Appleton.

aquamarine

A blue variety of the mineral beryl. A semiprecious gemstone, it is used in jewelry.

aquifer

A body of rock through which appreciable amounts of water can flow. The rock of an aquifer must be porous and permeable (full of interconnected holes) so that it can conduct water. Aquifers are an important source of fresh water, for example, for drinking and irrigation, in many arid areas of the world, and are exploited by the use of artesian wells. An aquifer may be underlain, overlain, or sandwiched between less permeable layers, called aquicludes or aquitards, which impede water movement. Sandstones and porous limestones make the best aquifers.

aragonite

A white, yellowish, or gray mineral, calcium carbonate, $CaCO_3$, a denser, harder polymorph (substance having the same chemical composition but different structure) of calcite. Secreted by corals, mollusks, and green algae, it is an important constituent of shallow marine muds and pearl.

arch

In geomorphology, any natural bridgelike land feature formed by erosion. A *sea arch* is formed from the wave erosion of a headland where the backs of two caves have met and broken through. The roof of the arch eventually collapses to leave part of the headland isolated in the sea as a stack. A *natural bridge* is formed on land by wind or water erosion and spans a valley or ravine.

Archean

Also *Archeozoic.* A widely used term for the earliest era of geological time; the first part of the Precambrian Eon, spanning the interval from the formation of Earth to about 2.5 billion years ago.

archipelago

A group of islands, or an area of sea containing a group of islands. The islands of an archipelago are usually volcanic in origin, and they sometimes represent the tops of peaks in areas around continental margins flooded by the sea. Volcanic islands are formed either when a hot spot within the earth's mantle produces a chain of volcanoes on the surface, such as the Hawaiian Archipelago or at a destructive plate margin (*see* **plate tectonics**) where the subduction of one plate beneath another produces an arc-shaped island group called an island arc, such as the Aleutian Archipelago. Novaya Zemlya in the Arctic Ocean, on the other hand, is the northern extension of the Ural Mountains and resulted from continental flooding.

Arctic Circle

An imaginary line that encircles the North Pole at latitude 66° 33' north. Within this line there is at least one day in the summer during which the sun never sets, and at least one day in the winter during which the sun never rises. The length of the periods of continuous daylight in summer and darkness in winter varies with the nearness to the North Pole; the nearer the Pole, the longer the period during which the sun is continually above or below the horizon.

arête

French for "fish-bone." A North American *combe-ridge,* which is a sharp narrow ridge separating two glacial troughs (valleys), or corries. The typical U-shaped cross sections of glacial troughs give arêtes very steep sides. Arêtes are common in glaciated mountain regions such as the Rockies, the Himalayas, and the Alps.

arid region

A region that is very dry and has little vegetation. Aridity depends on temperature, rainfall, and evaporation, and so is difficult to quantify, but an arid area is usually defined as one that receives less than 250 mm/10 in of rainfall each year. (By comparison, New York City receives 1,120 mm/44 in per year.) There are arid regions in North Africa, Pakistan, Australia, the United States, and elsewhere. Very arid regions are deserts.

The scarcity of fresh water is a problem for the inhabitants of arid zones, and constant research goes into discovering cheap methods of distilling sea water and artificially recharging natural groundwater reservoirs. Another problem is the eradication of salt in irrigation supplies from underground sources or from places where surface deposits form in poorly drained areas.

artesian well

A well that is supplied with water rising naturally from an underground water-saturated rock layer (aquifer). The water rises from the aquifer under its own pressure. Such a well may be drilled into an aquifer that is confined by impermeable rocks both above and below. If the water table (the top of the region of water saturation) in that aquifer is above the level of the well head, hydrostatic pressure will force the water to the surface.

Artesian wells are often overexploited because their water is fresh and easily available, and they eventually become unreliable. There is also some concern that pollutants such as pesticides or nitrates can seep into the aquifers. Much use is made of artesian wells in eastern Australia, where aquifers filled by water in the Great Dividing Range run beneath the arid surface of the Simpson Desert. The artesian well is named for Artois, a French province, where the phenomenon was first observed.

asbestos

Any of several related minerals of fibrous structure that offer great heat resistance because of their nonflammability and poor conductivity. Commercial asbestos is generally made either from serpentine (white asbestos) or from sodium iron silicate (blue asbestos). The fibers are woven together or bound by an inert material. Over time the fibers can work loose and, because they are small enough to float freely in the air or be inhaled, exposure to its dust can cause cancer. For this reason, asbestos usage is now strictly controlled.

Asbestos has been used for brake linings, suits for fire fighters and astronauts, insulation of electric wires in furnaces, and fireproof materials for the building industry. Exposure to asbestos is a recognized cause of industrial cancer (mesothelioma), especially in the blue form (from South Africa), rather than the more common white form. Asbestosis is a chronic lung inflammation caused by asbestos dust.

asphalt

A mineral mixture containing semisolid brown or black bitumen, used in the construction industry. Asphalt is mixed with rock chips to form paving material, and the purer varieties are used as insulating material and for waterproofing masonry. Asphalt can be produced artificially by the distillation of petroleum. Considerable natural deposits of asphalt occur around the Dead Sea and in the Philippines, Cuba, Venezuela, and Trinidad. Bituminous limestone occurs at Neufchâtel, France. The availability of recycled colored glass led in 1988 to the invention of *glassphalt,* asphalt that is 15 percent crushed glass. It is used to pave roads in New York.

asthenosphere

A layer within Earth's mantle lying beneath the lithosphere, typically beginning at a depth of approximately 100 km/63 mi and extending to depths of approximately 260 km/160 mi. Sometimes referred to as the weak sphere, it is characterized by being weaker and more elastic than the surrounding mantle. The asthenosphere's elastic behavior and low viscosity allow the overlying, more rigid plates of lithosphere to move laterally in a process known as plate tectonics. Its elasticity and viscosity also allow overlying crust and mantle to move vertically in response to gravity to achieve *isostatic equilibrium* (*see* **isostasy**).

atmosphere

The mixture of gases surrounding a planet. Planetary atmospheres are prevented from escaping by the pull of gravity. On Earth and other planets, atmospheric pressure decreases with altitude. In its lowest layer, Earth's atmosphere consists of nitrogen (78 percent) and oxygen (21 percent), both in molecular form (two atoms bonded together) and 1 percent argon. Small quantities of other gases are important to the chemistry and physics of the earth's atmosphere, including water and carbon dioxide. The atmosphere plays a major part in the various cycles of nature (the water cycle, the carbon cycle, and the nitrogen cycle). It is the principal industrial source of nitrogen, oxygen, and argon, which are obtained by fractional distillation of liquid air.

The earth's atmosphere is divided into four regions of atmosphere classified by temperature:

1. *troposphere.* This is the lowest level of the atmosphere (altitudes from 0 to 10 km/6 mi) and is heated to an average temperature of 15°C/59°F by the earth, which in turn is warmed by infrared and visible radiation from the sun. Warm air cools as it rises in the troposphere and this rising of warm air causes rain and most other weather phenomena. The top of the troposphere is approximately -60°C.
2. *stratosphere.* Temperature increases with altitude in this next layer (from 10 km/6 mi to 50 km/31 mi), from -60°C/-140°F to near 0°C/32°F.
3. *mesosphere.* Temperature again decreases with altitude through the mesosphere (50 km/31 mi to 80 km/50 mi), from 0°C/32°F to below -100°C/-212°F.
4. *thermosphere.* In the highest layer (80 km/50 mi to about 700 km/435 mi), temperature rises with altitude to extreme values of thousands of degrees. The meaning of these extreme temperatures can be misleading. High thermosphere temperatures represent little heat because they are defined by motions among so few atoms and molecules spaced widely apart from one another.

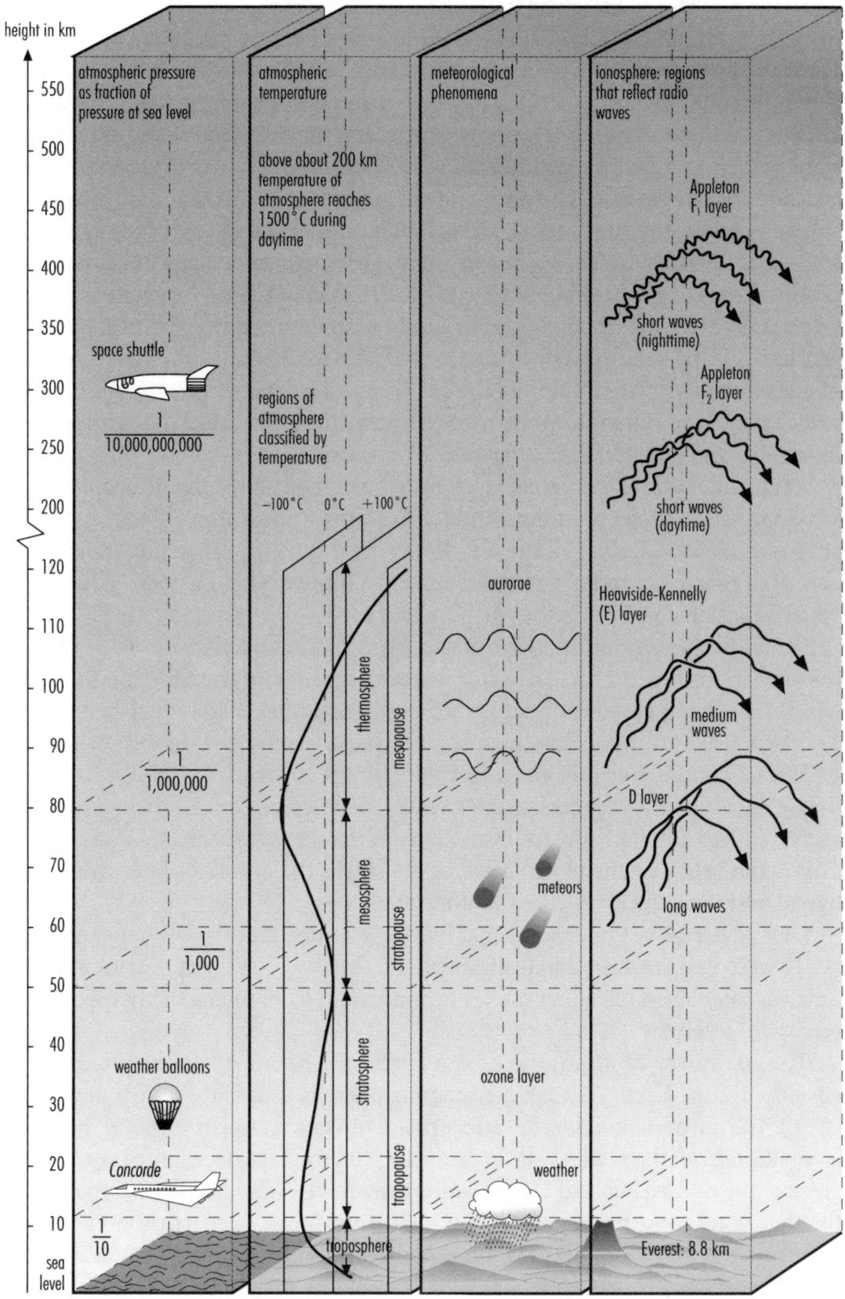
height in km
550
500
450
400
350
300
250
200
120
110
100
90
80
70
60
50
40
30
20
10
sea level
atmospheric pressure as fraction of pressure at sea level
space shuttle
1/10,000,000,000
1/1,000,000
1/1,000
weather balloons
Concorde
1/10
atmospheric temperature
above about 200 km temperature of atmosphere reaches 1500 °C during daytime
regions of atmosphere classified by temperature
−100 °C
0 °C
+100 °C
thermosphere
mesopause
mesosphere
stratopause
stratosphere
tropopause
troposphere
meteorological phenomena
aurorae
meteors
ozone layer
weather
ionosphere: regions that reflect radio waves
Appleton F1 layer
short waves (nighttime)
Appleton F2 layer
short waves (daytime)
Heaviside-Kennelly (E) layer
medium waves
D layer
long waves
Everest: 8.8 km

atmosphere

The thermal structure of the earth's atmosphere is the result of a complex interaction between the electromagnetic radiation from the sun, radiation reflected from the earth's surface, and molecules and atoms in the atmosphere. High in the thermosphere, temperatures are high because of collisions between ultraviolet (UV) photons and atoms of the atmosphere. Temperature decreases at lower levels because there are fewer UV photons available, having been absorbed by collisions higher up. The thermal minimum that results at the base of the thermosphere is called the mesopause. The temperature maximum near the top of the stratosphere is called the stratopause. Here, temperatures rise as UV photons are absorbed by heavier molecules to form new gases. An important example is the production of ozone molecules (oxygen atom triplets, O_3) from oxygen molecules. Ozone is a better absorber of ultraviolet radiation than ordinary (two-atom) oxygen, and it is the ozone layer within the stratosphere that prevents lethal amounts of ultraviolet light from reaching the earth's surface. The temperature minimum between the stratosphere and troposphere marks the influence of the earth's warming effects and is called the *tropopause.*

At altitudes above the ozone layer and above the base of the mesosphere (50 km/31 mi), ultraviolet photons collide with atoms, knocking out electrons to create a plasma of electrons and positively charged ions. The resulting ionosphere acts as a reflector of radio waves, enabling radio transmissions to "hop" between widely separated points on the earth's surface.

Far above the atmosphere lie the Van Allen radiation belts. These are regions in which high-energy charged particles traveling outward from the sun (the solar wind) have been captured by the earth's magnetic field. The outer belt (about 1,600 km/1,000 mi) contains mainly protons, the inner belt (about 2,000 km/1,250 mi) contains mainly electrons. Sometimes electrons spiral down toward the earth, noticeably at polar latitudes, where the magnetic field is strongest. When such particles collide with atoms and ions in the thermosphere, light is emitted. This is the origin of the glow visible in the sky as the aurora borealis (northern lights) and the aurora australis (southern lights).

During periods of intense solar activity, the atmosphere swells outward; there is a 10 to 20 percent variation in atmospheric density. One result is to increase the drag on satellites. This effect makes it impossible to predict exactly the time of reentry of satellites.

The chemistry of atmospheres is related to the geology of the planets they envelop. Unlike Earth, Venus's dense atmosphere is dominantly carbon dioxide (CO_2). The carbon dioxide–rich atmosphere of Venus absorbs infrared radiation emanating from the planet's surface, causing very high surface temperatures that are capable of melting lead (*see* **greenhouse effect**). If all of the carbon dioxide that has gone to form carbonate rock (*see* **limestone**) on Earth were liberated into the troposphere, our atmosphere would be similar to that of Venus. It is the existence of liquid water that enables carbonate rock to form on Earth that has caused our atmosphere to differ substantially from the Venusian atmosphere.

Other atmospheric ingredients are found in particular localities: gaseous compounds of sulfur and nitrogen in towns, salt over the oceans; and everywhere dust composed of inorganic particles, decaying organic matter, tiny seeds and pollen from plants, and bacteria. Of particular importance are the anthropogenic chlorofluorocarbons (CFC) that destroy stratospheric ozone.

atmospheric pressure

The pressure at any point on the earth's surface that is due to the weight of the column of air above it; it therefore decreases as altitude increases. At sea level the average pressure is 101 kilopascals (1,013 millibars, 760 mmHg, or 14.7 lb per sq in). Changes in atmospheric pressure, measured with a barometer, are used in weather forecasting. Areas of relatively high pressure are called anticyclones; areas of low pressure are called *depressions.*

atoll

A continuous or broken circle of coral reef and low coral islands surrounding a lagoon.

attrition

The process by which particles of rock being transported by rivers, winds, or seas are rounded and gradually reduced in size by being struck against one another. The rounding of particles is a good indication of how far they have been transported. This is particularly true for particles carried by rivers, which become more rounded as the distance downstream increases.

avalanche

From the French *avaler,* meaning *to swallow.* The fall or flow of a mass of snow and ice down a steep slope under the force of gravity. Avalanches occur because of the unstable nature of snow masses in mountain areas. Changes of temperature, sudden sound, or earth-borne vibrations may trigger an avalanche, particularly on slopes of more than 35°. The snow compacts into ice as it moves, and rocks may be carried along, adding to the damage caused.

Avalanches leave slide tracks, long gouges down the mountainside that can be up to 1 km/0.6 mi long and 100 m/330 ft wide. These slides have a similar beneficial effect on biodiversity as do forest fires, clearing the land of snow and mature mountain forest. This enables plants and shrubs that cannot grow in shade to recolonize and creates wildlife corridors. Avalanches are particularly hazardous in ski resort areas such as the French Alps. In 1991 a massive avalanche considerably altered the shape of Mount Cook in New Zealand.

backwash

The retreat of a wave that has broken on a beach. When a wave breaks, water rushes up the beach as swash and is then drawn back toward the sea as backwash.

badlands

A barren landscape cut by erosion into a maze of ravines, pinnacles, gullies, and sharp-edged ridges. Areas in South Dakota and Nebraska are examples. Badlands, which can be created by overgrazing, are so called because of their total lack of value for agriculture and their inaccessibility.

bar

A deposit of sand or silt formed in a river channel, or a long sandy ridge running parallel to a coastline that is submerged at high tide (*see* **coastal erosion**). Coastal bars can extend across estuaries to form bay bars. These bars are greatly affected by the beach cycle. The high tides and high waves of winter erode the beach and deposit the sand in bars offshore.

barite

Barium sulfate, $BaSO_4$, the most common mineral of barium. It is white or light colored, and has a comparatively high density (specific gravity 4.6); the latter property makes it useful in the production of high-density drilling muds (muds used to cool and lubricate drilling equipment). Barite occurs mainly in ore veins, where it is often found with calcite and with lead and zinc minerals. It crystallizes in the orthorhombic system and can form tabular crystals or radiating fibrous masses.

barometer

An instrument that measures atmospheric pressure as an indication of weather. Most often used are the *mercury barometer* and the *aneroid barometer.* In a *mercury barometer* a column of mercury in a glass tube, roughly 0.75 m/2.5 ft high (closed at one end, curved upward at the other), is balanced by the pressure of the atmosphere on the open end; any change in the height of the column reflects a change in pressure. In an *aneroid barometer,* a shallow cylindrical metal box containing a partial vacuum expands or contracts in response to changes in pressure.

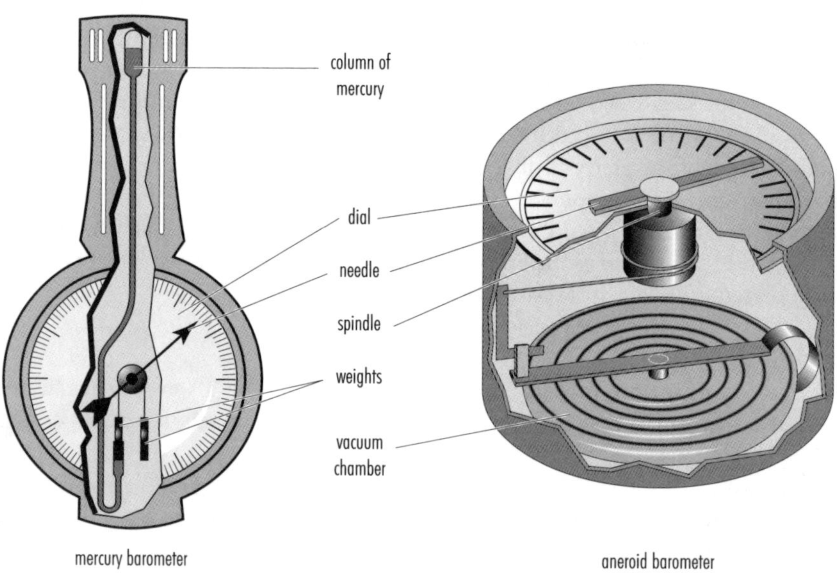

mercury barometer

aneroid barometer

barrier island

A long island of sand, lying offshore and parallel to the coast. Some are over 100 km/60 mi in length. Most barrier islands are derived from marine sands piled up by shallow longshore currents that sweep sand parallel to the seashore. Others are derived from former spits, connected to land and built up by drifted sand, that were later severed from the mainland. Often several islands lie in a continuous row offshore. Coney Island and Jones Beach near New York City are well-known examples, as is Padre Island, Texas. The Frisian Islands are barrier islands along the coast of the Netherlands.

barrier reef

A coral reef that lies offshore, separated from the mainland by a shallow lagoon.

basalt

The commonest volcanic igneous rock in the solar system. Much of the surfaces of the terrestrial planets Mercury, Venus, Earth, and Mars, as well as the moon, are composed of basalt. Earth's ocean floor is virtually entirely made of basalt. Basalt is mafic, that is, it contains relatively little silica: about 50 percent by weight. It is usually dark gray but can also be green, brown, or black. Its essential constituent minerals are calcium-rich feldspar and calcium- and magnesium-rich pyroxene.

The groundmass may be glassy or finely crystalline, sometimes with large crystals embedded. Basaltic lava tends to be runny and flows for great distances before solidifying. Successive eruptions of basalt have formed the great plateaus of Colorado and the Deccan plateau region of southwest India. In some places, such as Fingal's Cave in the Inner Hebrides of Scotland and the Giant's Causeway in Antrim, Northern Ireland, shrinkage during the solidification of the molten lava caused the formation of hexagonal columns.

The dark-colored lowland mare regions of the moon are underlain by basalt. Lunar mare basalts have higher concentrations of titanium and zirconium and lower concentrations of volatile elements like potassium and sodium relative to terrestrial basalts. Martian basalts are characterized by low ratios of iron to manganese relative to terrestrial basalts, as judged from some Martian meteorites (shergottites, a class of the SNC meteorites) and spacecraft analyses of rocks and soils on the Martian surface.

base level

The level, or altitude, at which a river reaches the sea or a lake. The river erodes down to this level. If base level falls (due to uplift or a drop in sea level), rejuvenation takes place.

baseflow

Also *groundwater flow.* The movement of water from land to rivers through rock. It is the slowest form of such water movement, and accounts for the constant flow of water in rivers during times of low rainfall, and makes up the river's base line on a hydrograph.

basic rock

Igneous rock with relatively low silica contents of 45 to 52 percent by weight, such as gabbro and basalt. Along with the terms *acid rock* and *ultrabasic rock* it is part of an outdated classification system based on the erroneous belief that silicon in rocks is in the form of silicic acid. Geologists today are more likely to use the descriptive term *mafic rock* or report the amount of SiO_2 in weight percent.

batholith

Large, irregular, deep-seated mass of intrusive igneous rock, usually granite, with an exposed surface of more than 100 sq km/40 sq mi. The mass forms by the intrusion or upswelling of magma (molten rock) through the surrounding rock. Batholiths form the core of some large mountain ranges like the Sierra Nevada of western North America.

According to plate tectonic theory, magma rises in subduction zones along continental margins where one plate sinks beneath another. The solidified magma becomes the central axis of a rising mountain range, resulting in the deformation (folding and overthrusting) of rocks on either side. Gravity measurements indicate that the downward extent or thickness of many batholiths is some 6–9 mi/10–15 km.

bathyal zone

A zone of the ocean, which lies on the continental slope. The bathyal zone begins at the continental margin, which is the top of the continental slope, at a depth of 200 m/650 ft, and extends to a depth of 2,000 m/6,500 ft. Above the bathyal zone on the continental shelf lies the sublittoral zone. Below the bathyal zone lies the abyssal zone. Bathyal zones (both temperate and tropical) have greater biodiversity than coral reefs, according to a 1995 study by the Natural History Museum in London.

bauxite

Principal ore of aluminum, consisting of a mixture of hydrated aluminum oxides and hydroxides, generally contaminated with compounds of iron, which give it a red color. It is formed by the chemical weathering of rocks in tropical climates. Chief producers of bauxite are Australia, Jamaica, Russia, Kazakhstan, Suriname, and Brazil. To extract aluminum from bauxite, high temperatures (about 800°C/1,470°F) are needed to make the ore molten. Strong electric currents are then passed through the molten ore. The process is only economical if cheap electricity is readily available, usually from a hydroelectric plant.

bayou

Corruption of French *boyau* for *gut*. Regional term for an oxbow lake or marshy offshoot of a river in the southern United States. Bayous may be formed, as in the lower Mississippi, by a river flowing in wide curves or meanders in flat country, and then cutting a straight course across them in times of flood, leaving loops of isolated water behind.

beach

A strip of land bordering the sea, normally consisting of boulders and pebbles on exposed coasts or sand on sheltered coasts. A beach is usually defined by the high- and low-water marks. A berm, a ridge of sand and pebbles, may be found at the farthest point inland that the water reaches.

The unconsolidated material of the beach consists of a rocky debris eroded from exposed rocks and headlands by the processes of coastal erosion, or material carried in by rivers. The material is transported to the beach, and along the beach, by longshore drift. Incoming waves (swash) hit the beach at an angle and carry sand onto the beach. Outgoing waves (backwash) draw back at right angles to the beach carrying sand with them. This zigzag pattern results in a net movement of the material in one particular direction along the coast.

When the energy of the waves decreases due to interaction with currents or changes in the coastline, more sand is deposited than is transported, building up to create depositional features such as spits, bars, and tombolos.

Attempts often are made to artificially halt longshore drift and increase deposition on a beach by erecting barriers (groynes) at right angles to the beach. These barriers cause sand to build up on their upstream side but deplete the beach on the downstream side, causing beach erosion. The finer sand can be moved about by the wind, forming sand dunes.

Beach erosion also occurs due to the natural seasonal beach cycle. Spring tides and the high waves of winter storms tend to carry sand away from the beach and deposit it offshore. In the summer, calmer waves and neap tides cause increased deposition of sand on the beach, known as *building.*

Apart from the natural process of longshore drift, a beach may be threatened by the commercial use of sand and aggregate by the mineral industry, because particles of metal ore are often concentrated into workable deposits by the wave action—and by pollution (for example, by oil spilled or dumped at sea).

bearing

The direction of a fixed point, or the path of a moving object, from a point of observation on the earth's surface, expressed as an angle from the north. Bearings are taken by compass and are measured in degrees (°), given as three-digit numbers increasing clockwise. For instance, north is 000°, northeast is 045°, south is 180°, and southwest is 225°.

True north differs slightly from magnetic north (the direction in which a compass needle points), hence northeast may be denoted as 045M or 045T, depending on whether the reference line is magnetic (M) or true (T) north. True north also differs slightly from grid north because it is impossible to show a spherical Earth on a flat map.

Beaufort scale

A system of recording wind velocity (speed), devised by Francis Beaufort in 1806. It is a numerical scale ranging from 0 to 17, calm being indicated by 0 and a hurricane by 12. A reading of 13 to 17 indicate degrees of hurricane force. In

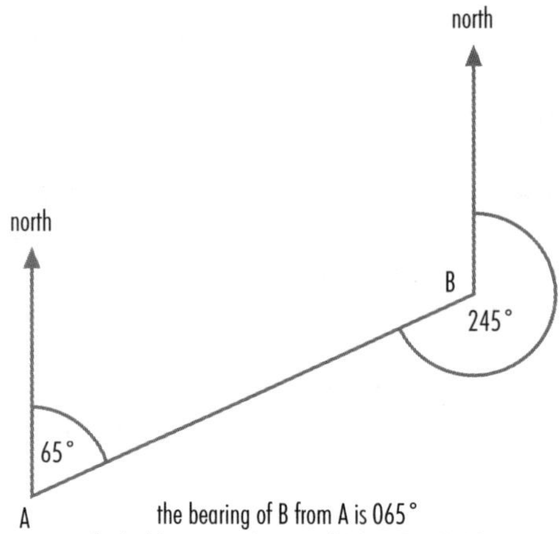

bearing

1874 the scale received international recognition; it was modified in 1926. Measurements are made at 10 m/33 ft above ground level.

bed

A single sedimentary rock unit with a distinct set of physical characteristics or contained fossils, readily distinguishable from those of beds above and below. Well-defined partings called *bedding planes* separate successive beds or strata. The depth of a bed can vary from a fraction of a centimeter to several meters or yards, and can extend over any area. The term is also used to indicate the floor beneath a body of water (lake bed) and a layer formed by a fall of particles (ash bed).

bedload

Material rolled or bounced (by saltation) along a river bed. The particles carried as bedload are much larger than those carried in suspension in the water. During a flood many heavy boulders may be moved in this way. Such boulders can be seen lying on the river bed during times of normal flow.

Benguela current

A cold ocean current in the South Atlantic Ocean, moving northward along the west coast of southern Africa and merging with the south equatorial current at a latitude of 15° S. Its rich plankton supports large, commercially exploited fish populations.

Benioff zone

A seismically active zone inclined from a deep sea trench beneath a continent or continental margin. Earthquakes along Benioff zones define the top surfaces of plates of lithosphere that descend into the mantle beneath another, overlying plate. The zone is named for Hugo Benioff, a U.S. seismologist who first described this feature.

benthic

Term describing the environment on the seafloor supporting bottom-dwelling plants, such as seaweeds, and animals, including corals, anemones, sponges, and shellfish.

bentonite

A soft porous rock consisting mainly of clay minerals, such as montmorillonite, and resembling fuller's earth, which swells when wet. It is formed by the chemical alteration of glassy volcanic material, such as tuff. Bentonite is used in paper making, molding sands, drilling muds for oil wells, and as a decolorant in food processing. On the Palos Verdes peninsula in southern California, bentonite layers contribute to landslides along the coast.

bergschrund

A deep crevasse that may be found at the head of a glacier. It is formed as a glacier pulls away from the headwall of the corrie, or hollow, in which it accumulated.

Beringia

Also *Bering Land Bridge.* A former land bridge, 1,600 km/1,000 mi wide, between Asia and North America; it existed during the ice ages that occurred before 35,000 B.C and during the period 24,000–9000 B.C. It is now covered by the Bering Strait and Chukchi Sea.

berm

On a beach, a ridge of sand or pebbles running parallel to the water's edge, formed by the action of the waves on beach material. Sand and pebbles are deposited at the farthest extent of swash (advance of water) to form a berm. Berms can also be formed well up a beach following a storm, when they are known as *storm berms.*

The United Nations coalition forces in the Gulf War (1991) adopted the term to mean a temporary defensive barrier formed by bulldozing the desert sand into a ridge.

beryl

A mineral, beryllium aluminum silicate, $Be_3Al_2Si_6O_{18}$, which forms crystals chiefly in granite. It is the chief ore of beryllium. Two of its gem forms are aquamarine (light-blue crystals) and emerald (dark-green crystals).

bight

A coastal indentation, crescent-shaped or gently curving, such as the Bight of Biafra in west Africa and the Great Australian Bight.

biogeochemistry

An emerging branch of geochemistry involving the study of how chemical elements and their isotopes move between living organisms and geological materials.

biological weathering

A form of weathering caused by the activities of living organisms—for example, the growth of roots or the burrowing of animals. Tree roots are probably the most significant agents of biological weathering, as they are capable of prying apart rocks by growing into cracks and joints.

biotite

Dark mica, $K(Mg, Fe)_3Al\ Si_3O_{10}(OH, F)_2$, a common silicate mineral. It is brown to black with shiny surfaces, and like all micas, it splits into very thin flakes along its one perfect cleavage. Biotite is a mineral found in igneous rocks, such as granites, and metamorphic rocks such as schists and gneisses.

bise

A cold dry northerly wind experienced in southern France and Switzerland.

bitumen

An impure mixture of hydrocarbons, including such deposits as petroleum, asphalt, and natural gas, although sometimes the term is restricted to a soft kind of pitch resembling asphalt. Solid bitumen may have arisen as a residue from the evaporation of petroleum. If evaporation took place from a pool or lake of petroleum, the residue might form a pitch or asphalt lake, such as Pitch Lake in Trinidad. Bitumen was used in ancient times as a mortar, and by the Egyptians for embalming.

black earth

Exceedingly fertile soil that covers a belt of land in northeast North America, Europe, and Asia. In Europe and Asia it extends from Bohemia through Hungary, Romania, southern Russia, and Siberia, as far as Manchuria, having been deposited when the great inland ice sheets melted at the close of the last ice age. In North America, it extends from the Great Lakes east through New York state, having been deposited when the last glaciers melted and retreated from the terminal moraine.

bluff

An alternative name for a river cliff.

bog

A type of wetland in which decomposition is slowed down and dead plant matter accumulates as peat. Bogs develop under conditions of low temperature, high acidity, low nutrient supply, stagnant water, and oxygen deficiency. Typical bog plants are sphagnum moss, rushes, and cotton grass; insectivorous plants such as sundews and bladderworts are common in bogs (insect prey make up for the lack of nutrients).

Large bogs are found in Ireland and northern Scotland. They are dominated by heather, cotton grass, and over 30 species of sphagnum mosses that make up the general bog matrix. The rolling blanket bogs of Scotland are a southern outcrop of the Arctic tundra ecosystem, and have taken thousands of years to develop.

bolson

A basin without an outlet, found in desert regions. Bolsons often contain temporary lakes, called *playa lakes,* and become filled with sediment from inflowing intermittent streams.

borax

Hydrous sodium borate, $Na_2B_4O_7 \cdot 10H_20$, found as soft whitish crystals or encrustations on the shores of hot springs and in the dry beds of salt lakes in arid regions, where it occurs with other borates, halite, and gypsum. It is used in bleaches and laundry detergents. A large industrial source is Borax Lake, California. Borax is also used in glazing pottery, in soldering, as a mild antiseptic, and as a metallurgical flux.

bore

A surge of tidal water up an estuary or a river, caused by the funneling of the rising tide in a narrowing river mouth. A very high tide, possibly fanned by wind, may build up when it is held back by a river current in the river mouth. The result is a broken wave, a meter or a few feet high, that rushes upstream. Famous bores are found in the rivers Severn (England), Seine (France), Hooghly (India), and Chang Jiang (China), where bores of over 4 m/13 ft in height have been reported.

Bouguer anomaly

An geophysics, an increase in the earth's gravity observed near a mountain or dense rock mass. This is due to the gravitational force exerted by the rock mass. It is named for its discoverer, the French mathematician Pierre Bouguer (1698–1758), who first observed it 1735.

boulder clay

Another name for *till,* a type of glacial deposit.

braiding

The subdivision of a river into several channels caused by deposition of sediment as islets in the channel. Braided channels are common in meltwater streams.

breccia

Coarse-grained clastic sedimentary rock, made up of broken fragments *(clasts)* of preexisting rocks held together in a fine-grained matrix. It is similar to conglomerate but the fragments in breccia are jagged in shape.

butte

A steep-sided, flat-topped hill, formed in horizontally layered sedimentary rocks, largely in arid areas. A large butte with a pronounced tablelike profile is a *mesa.* Buttes and mesas are characteristic of semiarid areas where remnants of resistant rock layers protect softer rock underneath, as in the plateau regions of Colorado, Utah, and Arizona.

calcite

A colorless, white, or light-colored common rock-forming mineral, calcium carbonate, $CaCO_3$. It is the main constituent of limestone and marble and forms many types of invertebrate shell. Calcite often forms stalactites and stalagmites in caves and is also found deposited in veins through many rocks because of the ease with which it is dissolved and transported by groundwater. Oolite is a rock consisting of spheroidal calcite grains. It rates 3 on the Mohs scale of hardness. Large crystals up to 1 m/3 ft have been found in Oklahoma and Missouri. Iceland spar is a transparent form of calcite used in the optical industry. As limestone it is used in the building industry.

caldera

A very large basin-shaped crater. Calderas are found at the tops of volcanoes, where the original peak has collapsed into an empty chamber beneath. The basin, many times larger than the original volcanic vent, may be flooded, producing a crater lake, or the flat floor may contain a number of small volcanic cones, produced by volcanic activity after the collapse. Typical calderas are Kilauea, Hawaii; Crater Lake, Oregon; and the summit of Olympus Mons, on Mars. Some calderas are wrongly referred to as craters, such as Ngorongoro, Tanzania.

California current

A cold ocean current in the eastern Pacific Ocean flowing southward down the western coast of North America. It is part of the North Pacific gyre (a vast, circular movement of ocean water).

calima

From the Spanish word for *haze,* a dust cloud in Europe, originating in the Sahara Desert. It sometimes causes heat waves and eye irritation.

Cambrian

The period of geological time 570 million to 510 million years ago; the first period of the Paleozoic era. All invertebrate animal life appeared at this time, and marine algae were widespread. The *Cambrian Explosion,* 530 million to 520 million years ago, saw the first appearance in the fossil record of all modern animal phyla. The earliest fossils with hard shells, such as trilobites, date from this period. The name comes from *Cambria,* the medieval Latin name for Wales, where Cambrian rocks are typically exposed and were first described.

Canaries current

A cold ocean current in the North Atlantic Ocean flowing southwest from Spain along the northwest coast of Africa. It meets the northern equatorial current at a latitude of 20° N.

canyon

From the Spanish *cañon,* meaning *tube.* A deep, narrow valley or gorge running through mountains. Canyons are formed by stream down cutting, usually in arid areas, where the rate of down cutting is greater than the rate of weathering, and where the stream or river receives water from outside the area. There are many canyons in the western United States and in Mexico, for example the Grand Canyon of the Colorado River in Arizona, the canyon in Yellowstone National Park, and the Black Canyon in Colorado.

carbon cycle

Circulation of carbon through the natural world. The movements of carbon from one reservoir to another on Earth (carbon fluxes) controls the amount of carbon dioxide in the atmosphere. Carbon dioxide is released into the atmosphere by

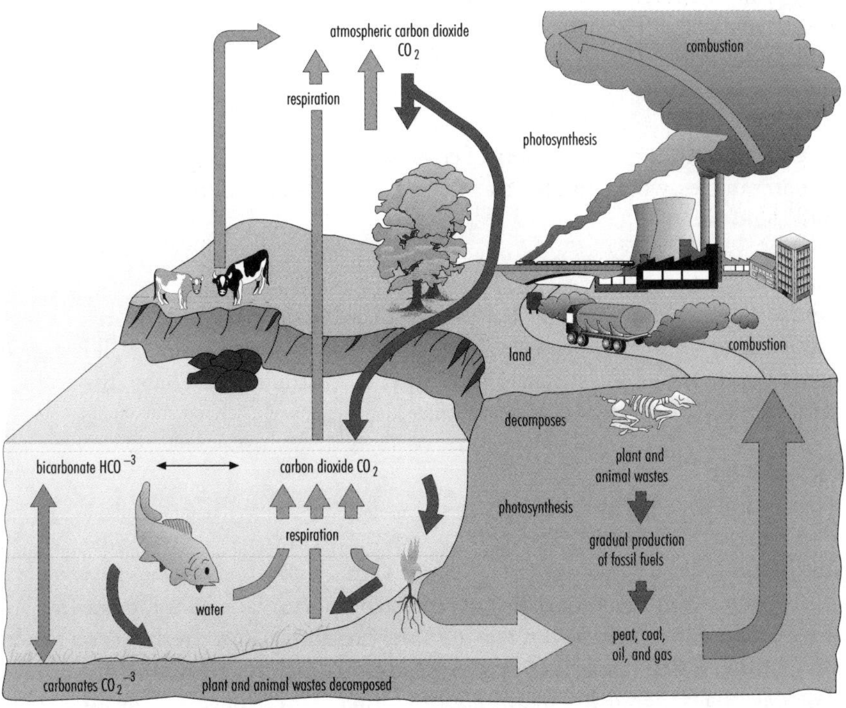

volcanism, metamorphism, decay of organic matter, burning of fossil fuels, and respiration by living organisms. It is drawn out of the atmosphere by chemical weathering of silicate rock, reactions between the atmosphere and the oceans (eventually forming carbonate rocks like limestone), burial of organic matter, and plant photosynthesis.

Today, the carbon cycle is in danger of being disrupted by the increased burning of fossil fuels and the burning of large tracts of tropical forests, both of which can release large amounts of carbon in to the atmosphere, contributing to the greenhouse effect.

carbonation

A form of chemical weathering caused by rainwater that has absorbed carbon dioxide from the atmosphere and formed a weak carbonic acid. The slightly acidic rainwater is then capable of dissolving certain minerals in rocks. Limestone is particularly vulnerable to this form of weathering.

Carboniferous

The period of geological time from 362.5 million to 290 million years ago, the fifth period of the Paleozoic era. In the United States it is divided into two periods: the Mississippian (lower) and the Pennsylvanian (upper). Typical of the lower-Carboniferous rocks are shallow-water limestones, while upper-Carboniferous rocks have delta deposits with coal (hence the name). Amphibians were abundant and reptiles evolved during this period.

carnelian

A semiprecious gemstone variety of chalcedony consisting of quartz (silica) with iron impurities, which give it a translucent red color. It is found mainly in Brazil, India, and Japan.

carnotite

Potassium uranium vanadate, $K_2(UO_2)_2(VO_4)_2.3H_2O$, a radioactive ore of vanadium and uranium with traces of radium. A yellow powdery mineral, it is mined chiefly in the Colorado Plateau; Radium Hill, Australia; and Shaba, the Democratic Republic of Congo (formerly Zaire).

cartography

The art and practice of drawing maps.

cassiterite

Also *tinstone.* A mineral consisting of reddish brown to black stannic oxide (SnO_2), usually found in granite rocks. It is the chief ore of tin. When fresh it has a bright adamantine luster. It was formerly extensively mined in Cornwall, England; today Malaysia is the world's main supplier. Other sources of cassiterite are Africa, Indonesia, and South America.

cataclastic rock

Metamorphic rock, such as a breccia, containing angular fragments of preexisting rock produced by the grinding and crushing action (cataclasis) of faults.

catchment area

In earth sciences, the area from which water is collected by a river and its tributaries. In the social sciences the term may be used to denote the area from which people travel to obtain a particular service or product, such as the area from which a school draws its pupils.

cave

A roofed-over cavity in the earth's crust usually produced by the action of underground water or by waves on a seacoast. Caves of the former type commonly occur in areas underlain by limestone, such as Kentucky and many Balkan regions, where the rocks are soluble in water. A *pothole* is a vertical hole in rock caused by water descending a crack; it is thus open to the sky.

Limestone caves. Most inland caves are found in karst regions, because limestone is soluble when exposed to ground water. As the water makes its way along the main joints, fissures, and bedding planes, they are constantly enlarged into potential cave passages, which ultimately join to form a complex network. Stalactites and stalagmites and columns form due to water rich in calcium carbonate dripping from the roof of the cave. The collapse of the roof of a cave produces features such as *natural arches* and *steep-sided gorges.* Limestone caves are usually found just below the water-table, wherever limestone outcrops on the surface. The biggest cave in the world is over 70 km/44 mi long at Holloch, Switzerland.

Sea caves. Coastal caves are formed where relatively soft rock or rock containing definite lines of weakness, like basalt at tide level, is exposed to severe wave action. The gouging process (corrasion) and dissolution (corrosion) of weaker, more soluble rock layers is exacerbated by subsidence, and the hollow in the cliff face grows still larger because of air compression in the chamber. Where the roof of a cave has fallen in, the vent up to the land surface is called a *blow-hole.* If this grows, finally destroying the cave form, the outlying truncated "portals" of the cave are known as *stacks* or *columns.* The Old Man of Hoy (137 m/450 ft high), in the Orkney Islands, is a fine example of a stack.

During the Ice Age humans began living in caves, leaving many layers of debris that archeologists have unearthed and dated in the Old World and the New. They also left cave art, paintings of extinct animals often with hunters on their trail.

Cave animals often show loss of pigmentation or sight, and under isolation, specialized species may develop. The scientific study of caves is called *speleology.*

Celebrated caves include the Mammoth Cave in Kentucky, 6.4 km/4 mi long and 38 m/125 ft high; the Caverns of Adelsberg (Postumia) near Trieste, Italy, which extend for many miles; Carlsbad Cave, New Mexico, the largest in the United States; the Cheddar Caves, England; Fingal's Cave, Scotland, which has a range of basalt columns; and Peak Cavern, England.

cavitation

Erosion of rocks caused by the forcing of air into cracks by hydraulic action. Cavitation results from the pounding of waves on the coast and the swirling of turbulent river currents, and exerts great pressure, eventually causing rocks to break apart. The process is particularly common at waterfalls, where the turbulent falling water contains many air bubbles, which burst to send shock waves into the rocks of the river bed and banks.

celestine

A mineral consisting of strontium sulfate, $SrSO_4$, occurring as white or light blue crystals, also called *celestite.* Celestine occurs in cavity linings associated with calcite, dolomite, or fluorite. It is the principal source of strontium. Celestine is found in small quantities in Germany, Italy, and the United States.

cement

Any bonding agent used to unite particles in a single mass or to cause one surface to adhere to another. *Portland cement* is a powder that, when mixed with water and sand or gravel, turns into mortar or concrete. In geology, cement refers to a chemically precipitated material such as carbonate that occupies the interstices of clastic rocks.

The term *cement* covers a variety of materials, such as fluxes and pastes, and also bituminous products obtained from tar. In 1824 English bricklayer Joseph Aspdin (1779–1855) created and patented the first portland cement, so named because its color in the hardened state resembled that of portland stone, a limestone used in building.

Cement is made by heating limestone (calcium carbonate) with clay (which contains a variety of silicates along with aluminum). This produces a gray powdery mixture of calcium and aluminum silicates. On addition of water, a complex series of reactions occurs and calcium hydroxide is produced. Cement sets by losing water.

Cement kilns contribute somewhat to the world's output of carbon dioxide. It was estimated in 1997 that by the year 2000 cement works could be responsible for 10 percent of all emissions of carbon dioxide. Cement production is growing fastest in East Asia, with rapid construction of buildings and roads.

Cenozoic

Also *Caenozoic.* The era of geological time that began 65 million years ago and continues to the present day. It is divided into the Tertiary and Quaternary periods. The Cenozoic marks the emergence of mammals as a dominant group, including humans, and the formation of the mountain chains of the Himalayas and the Alps.

chalcedony

A form of the mineral quartz, SiO_2, in which the crystals are so fine grained that they are impossible to distinguish with a microscope (cryptocrystalline). Agate, onyx, and carnelian are gem varieties of chalcedony.

chalcopyrite

A copper iron sulfide mineral, $CuFeS_2$, the most common ore of copper. It is brassy yellow in color and may have an iridescent surface tarnish. It occurs in many different types of mineral vein, in rocks ranging from basalt to limestone.

chalk

A soft, fine-grained, whitish sedimentary rock composed of calcium carbonate, $CaCO_3$, extensively quarried for use in cement, lime, and mortar, and in the manufacture of cosmetics and toothpaste. Blackboard chalk in fact consists of gypsum (calcium sulfate, $CaSO_4.2H_2O$).

Chalk was once thought to derive from the remains of microscopic animals, or foraminifera. In 1953, however, it was seen under the electron microscope to be composed chiefly of coccolithophores, unicellular lime-secreting algae, and hence primarily of plant origin. It is formed from deposits of deep-sea sediments called *oozes.*

Chalk was laid down in the later Cretaceous period and covers a wide area in Europe. In England it stretches in a belt from Wiltshire and Dorset continuously across Buckinghamshire and Cambridgeshire to Lincolnshire and Yorkshire, and also forms the North and South Downs, and the cliffs of southern and southeastern England.

channel efficiency

The ability of a river channel to discharge water. Channel efficiency can be assessed by calculating the channel's hydraulic radius. The most efficient channels are generally semicircular in cross section, and it is this shape that water engineers try to create when altering a river channel to reduce the risk of flooding.

chaparral

The thick scrub country of the southwestern United States. Thorny bushes have replaced what was largely evergreen live oak trees.

chemical weathering

A form of weathering brought about by chemical attack of rocks, usually in the presence of water. Chemical weathering involves the "rotting," or breakdown, of the original minerals within a rock to produce new minerals (such as clay minerals, bauxite, and calcite). Some chemicals are dissolved and carried away from the weathering source.

A number of processes bring about chemical weathering, such as carbonation (breakdown by weakly acidic rainwater), hydrolysis (breakdown by water), hydration (breakdown by the absorption of water), and oxidation (breakdown by the oxygen in water). The reaction of carbon dioxide gas in the atmosphere with silicate minerals in rocks to produce carbonate minerals (*see* **calcite**) is called the *Urey reaction* after the chemist who proposed it, Harold Urey. The Urey reaction is an important link between Earth's climate and the geology of the planet. It has been proposed that chemical weathering of large mountain ranges like the

Himalayas of Nepal can remove carbon dioxide from the atmosphere by the Urey reaction (or other more complicated reactions like it), leading to a cooler climate as the greenhouse effects of the lost carbon dioxide are diminished.

china clay

See **kaolin.**

chinook

From the Native American tribe of the same name. The term means "snow-eater," a warm dry wind that blows downhill on the eastern side of the Rocky Mountains of North America. It often occurs in winter and spring, when it produces a rapid thaw, and so is important to the agriculture of the area. The chinook is similar to the föhn in the valleys of the European Alps.

chromite

Iron chromium oxide, $FeCr_2O_4$, the main chromium ore. It is one of the spinel group of minerals, and crystallizes in dark-colored octahedra of the cubic system. Chromite is usually found in association with ultramafic and mafic rocks; in Cyprus, for example, it occurs with serpentine, and in South Africa it forms continuous layers in a layered intrusion.

chromium ore

Essentially the mineral chromite, $FeCr_2O_4$, from which chromium is extracted. South Africa and Zimbabwe are major producers.

chrysotile

A mineral in the serpentine group, $Mg_3Si_20_5(OH)_4$. A soft, fibrous, silky mineral, it is the primary source of asbestos.

cinnabar

A mercuric sulfide mineral, HgS, the only commercially useful ore of mercury. It is deposited in veins and impregnations near recent volcanic rocks and hot springs. The mineral itself is used as a red pigment, commonly known as *vermilion.* Cinnabar is found in California, Spain (Almadén), Peru, Italy, and Slovenia.

cirque

French name for a corrie, a steep-sided hollow in a mountainside.

clay

A very fine-grained sedimentary deposit that has undergone a greater or lesser degree of consolidation. When moistened it is plastic, and it hardens on heating, which renders it impermeable. It may be white, gray, red, yellow, blue, or black, depending on its composition. Clay minerals consist largely of hydrous silicates of aluminum and magnesium together with iron, potassium, sodium, and organic substances. The crystals of clay minerals have a layered structure, capable of holding water, and are responsible for its plastic properties. According to international classification, in mechanical analysis of soil, clay has a grain size of less than 0.002 mm/0.00008 in.

Types of clay include adobe, alluvial clay, building clay, brick, cement, china clay (or kaolinite), ferruginous clay, fireclay, fusible clay, puddle clay, refractory clay, and vitrifiable clay. Clays have a variety of uses, some of which, such as pottery and bricks, date back to prehistoric times.

clay mineral

One of a group of hydrous silicate minerals that form most of the fine-grained particles in clays. Clay minerals are normally formed by weathering or alteration of other silicate minerals. Virtually all have sheet silicate structures similar to the micas. They exhibit the following useful properties: loss of water on heating; swelling and shrinking in different conditions; cation exchange with other media; and plasticity when wet. Examples are kaolinite, illite, and montmorillonite.

Kaolinite $Al_2Si_2O_5(OH)_4$ is a common white clay mineral derived from the alteration of aluminum silicates, especially feldspars. Illite contains the same constituents as kaolinite, plus potassium, and is the main mineral of clay sediments, mudstones, and shales; it is a weathering product of feldspars and other silicates. Montmorillonite contains the constituents of kaolinite plus sodium and magnesium; along with related magnesium- and iron-bearing clay minerals, it is derived from alteration and weathering of mafic igneous rocks. Kaolinite (the mineral name for kaolin or china clay) is economically important in the ceramic and paper industries. Illite, along with other clay minerals, may also be used in ceramics. Montmorillonite is the chief constituent of fuller's earth, and is also used in drilling muds (muds used to cool and lubricate drilling equipment). Vermiculite (similar to montmorillonite) will expand on heating to produce a material used in insulation.

climate

A combination of weather conditions at a particular place over a period of time—usually a minimum of 30 years. A classification of climate encompasses the averages, extremes, and frequencies of all meteorological elements such as temperature, atmospheric pressure, precipitation, wind, humidity, and sunshine, together with the factors that influence them. The primary factors involved are: the earth's rotation and latitudinal effects; ocean currents; large-scale movements of wind belts and air masses over the earth's surface; temperature differences between land and sea surfaces; and topography. Climatology, the scientific study of climate, includes the construction of computer-generated models, and considers not only present-day climates, their effects and their classification, but also long-term climate changes, covering both past climates (paleoclimates) and future predictions. Climatologists are especially concerned with the influence of human activity on climate change, among the most important of which, at both a local and global level, are those currently linked with ozone depleters and the greenhouse effect.

Climate classification. The word *climate* comes from the Greek *klima,* meaning an inclination or slope (referring to the angle of the sun's rays, and thus latitude). The earliest known classification of climate was that of the ancient Greeks, who based their system on latitudes. In recent times many different systems of classifying climate have been devised, most of which follow that formulated by the German climatologist Wladimir Köppen (1846–1940) in 1900. These systems use vegetation-based classifications such as desert, tundra, and rainforest. Classification by air mass is used in conjunction with this method. This idea was first introduced in 1928 by the Norwegian meteorologist Tor Bergeron, and links the climate of an area with the movement of the air masses it experiences.

In the eighteenth century, the British scientist George Hadley developed a model of the general circulation of atmosphere based on convection. He proposed a simple pattern of cells of warm air rising at the equator and descending at the Poles, known as *Hadley cells.* In fact, due to the rotation of the earth, there are three such cells in each hemisphere. The first two of these consist of air that rises at the equator and sinks at latitudes north and south of the tropics; the second two exist at the midlatitudes where the rising air from the subtropics flows toward the cold air masses of the third pair of cells circulating from the two polar regions. Thus, in this model, there are six main circulating cells of air above ground, producing seven terrestrial zones. These include three rainy regions (at the equator and the temperate latitudes) resulting from the moisture-laden rising air (*see* **intertropical convergence zone**), interspersed and bounded by four dry or desert regions (at the Poles and subtropics) resulting from the dry descending air.

Prevailing winds. Regions are also affected by different wind systems, which result from the rotation of the earth and the uneven heating of surface air. As air is heated by radiation from the sun, it expands and rises, and cooler air flows in to take its place. This movement of air produces belts of prevailing winds. Because of the rotation of the earth, these are deflected to the right in the Northern Hemisphere and to the left in the Southern Hemisphere. This effect, which is greater in the higher latitudes, is known as the *Coriolis effect.* Because winds transport heat and moisture, they affect the temperature, humidity, precipitation, and cloudiness of an area. As a result, regions with different prevailing wind directions may have different climates.

Temperature variations. The amount of heat received by the earth from the sun varies with latitude and season. In equatorial regions, there is no large seasonal variation in the mean daily temperature of the air near the ground, while in the polar regions, temperatures during the long winters when there is little incoming solar radiation fall far below summer temperatures. The temperature of the sea, and of the air above it, varies little in the course of day or night, whereas the surface of the land is rapidly cooled by lack of solar radiation. Similarly, the annual change of temperature is relatively small over the sea but much greater over the land. Thus, continental areas are colder than maritime regions in winter, but warmer in summer. This results in winds blowing from the sea that, relative to the land, are warm in winter and cool in summer, while winds originating from the central parts of continents are hot in summer and cold in winter. On average, air temperature drops with increasing land height at a rate of 1°C/1.8°F per 90 m/300 ft, so that even in equatorial regions, the tops of mountains can be snow covered throughout the year.

Vegetation-based climates. Rainfall is produced by the condensation of water vapor in air. When winds blow against a range of mountains the air is forced to ascend, resulting in precipitation (rain or snow), the amount depending on the height of the ground and the humidity of the air. The varied distribution of land and sea areas produces the complexity of the general circulation of the atmosphere; this, in turn, directly affects the distribution of climate. Centered on the equator is a belt of tropical rainforest, which may be either constantly wet or monsoonal (having wet and dry seasons in each year). On either side of this is a belt of savannah, with lighter seasonal rainfall and less dense vegetation, largely in the form of grasses. Then there is usually a transition through steppe (semiarid) to desert (arid), with a further transition through steppe to what is termed Mediterranean climate with dry summers. Beyond this is the moist temperate climate of middle latitudes, and then a zone of cold climate with moist winters. Where the desert extends into middle latitudes, however, the zones of Mediterranean and moist temperate climates are missing, and the transition is from desert to a cold climate with moist winters. In the extreme east of Asia a cold climate with dry winters extends from about 70° N to 35° N. The polar caps have tundra and glacial climates, with little or no precipitation.

Climate changes. Changes in climate can occur naturally or as a result of human activity. Natural variations can be caused by fluctuations in the amount of solar radiation reaching the earth—for example, sunspot activity is thought to produce changes in the earth's climate. Another theory, known as the *Milankovitch hypothesis,* holds that variations in the earth's orbit around the sun also brings about climatic changes. Natural events on the surface of the earth, such as volcanic eruptions and the effects of El Niño, can result in temporary climate changes on a worldwide scale, sometimes extending over several months or even years. Human influences on the climate range from localized effects such as cloud seeding to produce rain, to the global effects of acid rain from industrial emissions, pollution, desertification, and the destruction of the rainforests. The study of past climates, paleoclimatogy, involves the investigation of climate changes from the ice ages to the beginning of instrumental recording in the nineteenth century.

climate model

A computer simulation, based on physical and mathematical data, of the entire climatic system of the earth. It is used by researchers to study such topics as the possible long-term disruptive effects of the greenhouse gases (carbon dioxide, methane, and chlorofluorocarbons) or of variations in the amount of radiation given off by the sun.

climatic change

Change in the climate of an area or of the whole world over an appreciable period of time.

climatology

The study of climate, its global variations and causes.

Climatologists record mean daily, monthly, and annual temperatures and monthly and annual rainfall totals, as well as maximum and minimum values. Other data collected relate to pressure, humidity, sunshine, cold cover, and the frequency of days of frost, snow, hail, thunderstorms, and gales. The main facts are summarized in tables and climatological atlases published by nearly all the national meteorological services of the world.

climograph

A diagram that shows both the average monthly temperature and precipitation of a place.

clint

One of a number of flat-topped limestone blocks that make up a limestone pavement, the naturally occurring surface of some limestone due to weathering. Clints are separated from each other by enlarged joints called *grikes.*

cloud

Water vapor condensed into minute water particles that float in masses in the atmosphere. Clouds, like fogs or mists, which occur at lower levels, are formed by the cooling of air containing water vapor, which generally condenses around tiny dust particles.

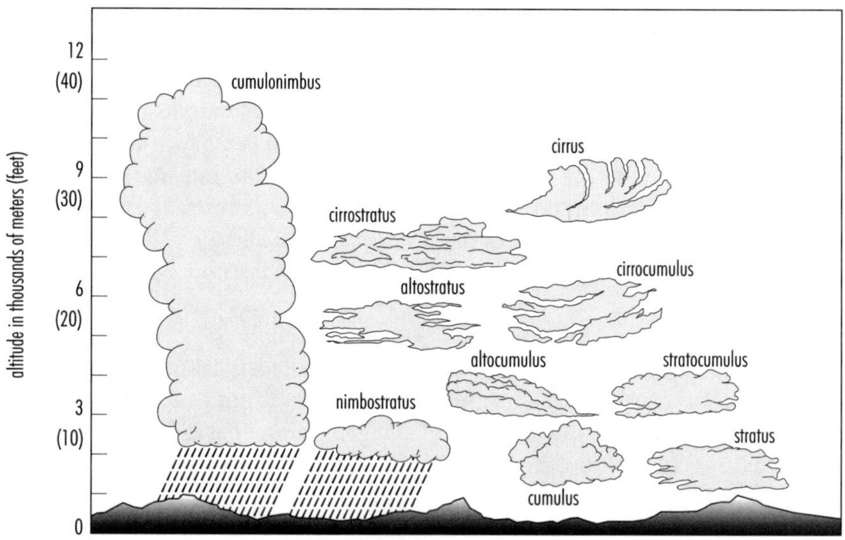

Clouds are classified according to the height at which they occur and their shape. *Cirrus* and *cirrostratus* clouds occur at around 10 km/33,000 ft. The former, sometimes called mares' tails, consist of minute specks of ice and appear as feathery white wisps, while cirrostratus clouds stretch across the sky as thin white sheets. Three types of cloud are found at 3–7 km/10,000–23,000 ft: cirrocumulus, altocumulus, and altostratus. *Cirrocumulus* clouds occur in small or large rounded tufts, sometimes arranged in the pattern called mackerel sky. *Altocumulus* clouds are similar, but larger, white clouds, also arranged in lines. *Altostratus* clouds are like heavy cirrostratus clouds and may stretch across the sky as a gray sheet. *Stratocumulus* clouds are generally lower, occurring at 2–6 km/6,500–20,000 ft. They are dull gray clouds that give rise to a leaden sky that may not yield rain. Two types of clouds, *cumulus* and *cumulonimbus,* are placed in a special category because they are produced by daily ascending air currents, which take moisture into the cooler regions of the atmosphere. Cumulus clouds have a flat base generally at 1.4 km/4,500 ft where condensation begins, while the upper part is dome-shaped and extends to about 1.8 km/6,000 ft. Cumulonimbus clouds have their base at much the same level, but extend much higher, often up to over 6 km/20,000 ft. Short heavy showers and sometimes thunder may accompany them. *Stratus* clouds, occurring below 1–2.5 km/3,000–8,000 ft, have the appearance of sheets parallel to the horizon and are like high fogs.

In addition to their essential role in the water cycle, clouds are important in the regulation of radiation in the earth's atmosphere. They reflect short-wave radiation from the sun, and absorb and reemit long-wave radiation from the earth's surface.

coal

A black or blackish mineral substance formed from the compaction of ancient plant matter in tropical swamp conditions. It is used as a fuel and in the chemical industry. Coal is classified according to the proportion of carbon it contains. The main types are anthracite (shiny, with about 90 percent carbon), bituminous coal (shiny and dull patches, about 75 percent carbon), and lignite (woody, grading into peat, about 50 percent carbon). Coal burning is one of the main causes of acid rain.

coastal erosion

The erosion of the land by the constant battering of the sea's waves, primarily by the processes of hydraulic action, corrasion, attrition, and corrosion. Hydraulic action occurs when the force of the waves compresses air pockets in coastal rocks and cliffs. The air expands explosively, breaking the rocks apart. Rocks and pebbles flung by waves against the cliff face wear it away by the process of corrasion. Chalk and limestone coasts are often broken down by dissolution (also called corrosion). Attrition is the process by which the eroded rock particles themselves are worn down, becoming smaller and more rounded.

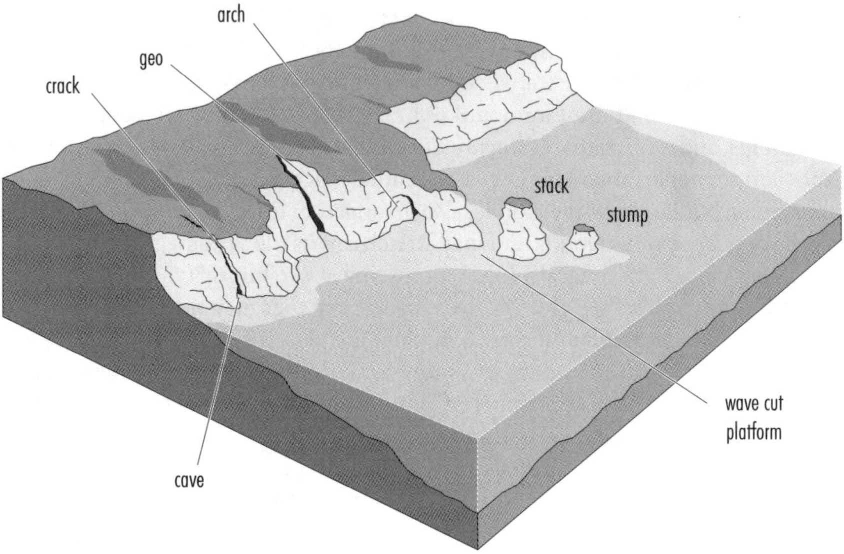

coastal erosion

Frost shattering (or freeze-thaw), caused by the expansion of frozen sea water in cavities, and biological weathering, caused by the burrowing of rock-boring mollusks, also result in the breakdown of the coastal rock.

Where resistant rocks form headlands, the sea erodes the coast in successive stages. First it exploits weaknesses, such as faults and cracks, in cave openings and then gradually wears away the interiors of the caves until their roofs are pierced through to form blowholes. In time, caves at either side of a headland may unite to form a natural arch. When the roof of the arch collapses, a stack is formed. This may be worn down further to produce a stump and a wave-cut platform.

Beach erosion occurs when more sand is eroded and carried away from the beach than is deposited by longshore drift. Beach erosion can occur due to the construction of artificial barriers, such as groynes, or due to the natural periodicity of the *beach cycle,* whereby high tides and the high waves of winter storms tend to carry sand away from the beach and deposit it offshore in the form of bars. During the calmer summer season some of this sand is redeposited on the beach.

coastal protection

Measures taken to prevent coastal erosion. Many stretches of coastline are so severely affected by erosion that beaches are swept away, threatening the livelihood of seaside resorts and causing buildings on the shore to become unsafe.

To reduce erosion, several different forms of coastal protection may be employed. Structures such as sea walls attempt to prevent waves reaching the cliffs by deflecting them back to sea. Such structures are expensive and of limited success. Adding sediment (beach nourishment) to make a beach wider causes waves to break early so that they have less power when they reach the cliffs. Wooden or concrete barriers called *groynes* may also be constructed at right angles to the beach in order to block the movement of sand along the beach (longshore drift).

cobalt ore

Cobalt is extracted from a number of minerals, the main ones being smaltite, $(CoNi)As_3$; linnaeite, Co_3S_4; cobaltite, CoAsS; and glaucodot, (CoFe)AsS. All commercial cobalt is obtained as a by-product of other metals, usually associated with other ores, such as copper. The Democratic Republic of Congo (formerly Zaire) is the largest producer of cobalt, and it is obtained there as a by-product of the copper industry. Other producers include Canada and Morocco. Cobalt is also found in the manganese nodules that occur on the ocean floor, and was successfully refined in 1988 from the Pacific Ocean nodules, although this process has yet to prove economic.

coccolithophorid

A microscopic, planktonic marine alga, which secretes a calcite shell. The shells (coccoliths) of coccolithophores are a major component of deep-sea ooze. Coccolithophores were particularly abundant during the late Cretaceous period and their remains form the northern European chalk deposits, such as the white cliffs of Dover.

combe

Also *coombe.* A steep-sided valley found on the scarp slope of a chalk escarpment. The inclusion of *combe* in a place name usually indicates that the underlying rock is chalk.

composite volcano

A steep-sided conical volcano formed above a subduction zone. Also called a *stratovolcano,* it is made up of alternate layers of ash and lava. The magma (molten rock) associated with composite volcanoes is very thick and often clogs up the vent. This can cause a tremendous buildup of pressure, which, once released, causes a very violent eruption. Examples of composite volcanoes are Mount St. Helens in Washington state and Mount Mayon in the Philippines. Compare the composite volcano with the shield volcano.

condensation

The conversion of a vapor to a liquid. This is frequently achieved by letting the vapor come into contact with a cold surface. It is the process by which water vapor turns into fine water droplets to form clouds.

Condensation in the atmosphere occurs when the air becomes completely saturated and is unable to hold any more water vapor. As the air rises, it cools and contracts. The cooler it becomes, the less water it can hold. Rain is frequently associated with warm weather fronts because the air rises and cools, allowing the water vapor to condense as rain. The temperature at which the air becomes saturated is known as the *dew point.* Water vapor does not condense in air if there are not enough condensation nuclei (particles of dust, smoke, or salt) for the droplets to form on. It is then said to be supersaturated. Condensation is an important part of the water cycle.

confluence
The point at which two rivers join.

conglomerate
Coarse-grained clastic sedimentary rock, composed of rounded fragments (clasts) of preexisting rocks cemented in a finer matrix, usually sand. The fragments in conglomerates are pebble- to boulder-sized, and the rock can be regarded as the lithified equivalent of gravel. A bed of conglomerate is often associated with a break in a sequence of rock beds (an *unconformity*), where it marks the advance of the sea over an old eroded landscape. An *oligomict conglomerate* contains one type of pebble; a *polymict conglomerate* has a mixture of pebble types. If the rock fragments are angular, it is called a *breccia.*

conservative margin
In plate tectonics, a region on the earth's surface in which one plate slides past another. An example is the San Andreas fault in California, where the movement of the plates is irregular and sometimes takes the form of sudden jerks, which cause the earthquakes common in the San Francisco and Los Angeles areas.

constructive margin
In plate tectonics, a region in which two plates are moving away from each other. It is also called a *divergent margin.* Magma, or molten rock, escapes to the surface along this margin to form new crust, usually in the form of a ridge. Over time, as more and more magma reaches the surface, the seafloor spreads—for example, the upwelling of magma at the Mid-Atlantic Ridge causes the floor of the Atlantic Ocean to grow at a rate of about 5 cm/2 in a year.

Volcanoes can form along the ridge and islands may result (for example, Iceland was formed in this way). Eruptions at constructive plate margins tend to be relatively gentle; the lava produced cools to form basalt.

continent
Any one of the seven large land masses of the earth, as distinct from the oceans. They are Asia, Africa, North America, South America, Europe, Australia, and Antarctica. Continents are constantly moving and evolving (*see* **plate tectonics**). A continent does not end at the coastline; its boundary is the edge of the shallow continental shelf, which may extend several hundred kilometers out to sea.

At the center of each continental mass lies a *shield* or *craton,* a deformed mass of old metamorphic rocks dating from Precambrian times. The shield is thick, compact, and solid (the Canadian Shield is an example), having undergone all the mountain-building activity it is ever likely to, and is usually worn flat. Around the shield is a concentric pattern of mountains comprised of folded rock, with older ranges, such as the Rockies, closest to the shield, and younger ranges, such as the coastal ranges of North America, farther away. This general concentric pattern is modified when two continental masses have drifted together and they become welded with a great mountain range along the joint, the way Europe and northern Asia are joined along the Urals. If a continent is torn apart, the new continental edges have no mountains formed by wrinkling and folding of the crust; for instance, the western coast of Africa, which drifted apart from South America 200 million years ago.

continental drift

The theory that, about 250 million to 200 million years ago, the earth consisted of a single large continent (Pangaea) that subsequently broke apart to form the continents known today. The theory was proposed in 1912 by German meteorologist Alfred Wegener, but such vast continental movements could not be satisfactorily explained until the theory of plate tectonics in the 1960s. The term *continental drift* is not strictly correct, because land masses do not drift through the oceans. The continents form part of a plate, and the amount of crust created at divergent plate margins must equal the amount of crust destroyed at subduction zones.

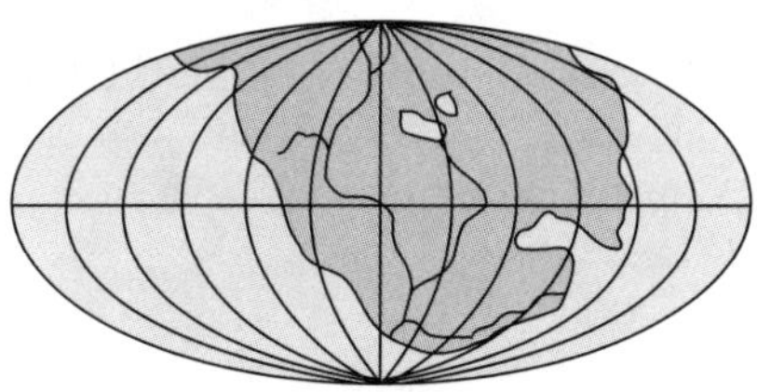

Upper Carboniferous period

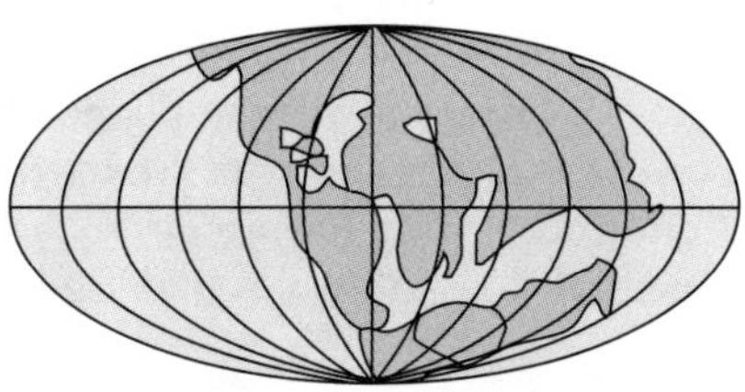

Eocene

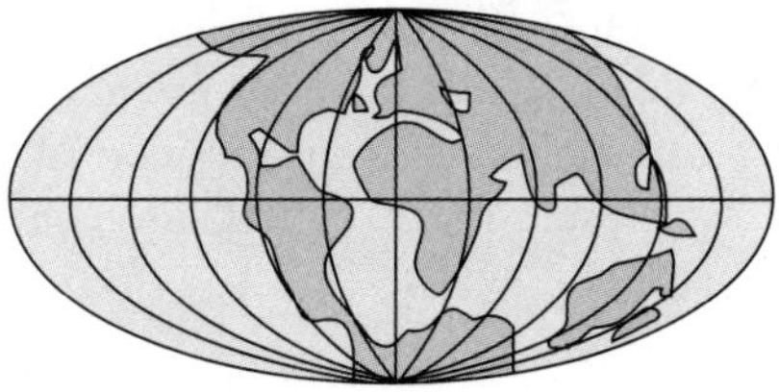

Lower Quaternary

continental rise

The portion of the ocean floor rising gently from the abyssal plain toward the steeper continental slope. The continental rise is a depositional feature formed from sediments transported down the slope, mainly by turbidity currents. Much of the continental rise consists of coalescing submarine alluvial fans bordering the continental slope.

continental shelf

The submerged edge of a continent, a gently sloping plain that extends into the ocean. It typically has a gradient of less than 1°. When the angle of the sea bed increases to between 1° and 5° (usually several hundred kilometers away from land), it becomes known as the *continental slope.*

continental slope

The sloping, submarine portion of a continent that extends downward from the edge of the continental shelf. In some places, such as south of the Aleutian Islands of Alaska, continental slopes extend directly to the ocean deeps or abyssal plain. In others, such as the east coast of North America, they grade into the gentler continental rises that in turn grade into the abyssal plains.

contour

On a map, a line drawn to join points of equal height. Contours are drawn at regular height intervals; for example, every 10 m/6.2 ft. The closer together the lines are, the steeper the slope. Contour patterns can be used to interpret the relief of an area and to identify land forms.

convection current

A current caused by the expansion of a liquid or gas as its temperature rises. The expanded material, being less dense, rises above colder and therefore denser material. Convection currents arise in the atmosphere above warm land masses or seas, giving rise to sea breezes and land breezes, respectively. In some heating systems, convection currents are used to carry hot water upward in pipes. Convection currents in the viscous rock of the earth's mantle help to drive the movement of the rigid plates making up the earth's surface (*see* **plate tectonics**).

convectional rainfall

Rainfall associated with hot climates, resulting from the uprising of convection currents of warm air. Air that has been warmed by the extreme heating of the ground surface rises to great heights and is abruptly cooled. The water vapor carried by the air condenses and rain falls heavily. Convectional rainfall is usually associated with a thunderstorm.

convergent margin

See **destructive margin.**

copper ore

Any mineral from which copper is extracted, including native copper, Cu; chalcocite, Cu_2S; chalcopyrite, $CuFeS_2$; bornite, Cu_5FeS_4; azurite, $Cu_3(CO_3)_2(OH)_2$; malachite, $Cu_2CO_3(OH)_2$; and chrysocolla, $CuSiO_3.2H_2O$. Native copper and the copper sulfides are usually found in veins associated with igneous intrusions. Chrysocolla and the carbonates are products of the weathering of copper-bearing rocks. Copper was one of the first metals to be worked, because it occurred in native form and needed little refining. Today the main producers are the United States, Russia, Kazakhstan, Georgia, Uzbekistan, Armenia, Zambia, Chile, Peru, Canada, and the Democratic Republic of Congo (formerly Zaire).

coral, coral reef

A marine invertebrate of the class Anthozoa in the phylum Cnidaria, which also includes sea anemones and jellyfish. It has a skeleton of lime (calcium carbonate) extracted from the surrounding water. Corals exist in warm seas, at moderate depths with sufficient light. Some coral is valued for decoration or jewelry, for example, Mediterranean red coral *Corallum rubrum.*

Corals live in a symbiotic relationship with microscopic algae (zooxanthellae), which are incorporated into the soft tissue. The algae obtain carbon dioxide from the coral polyps, and the polyps receive nutrients from the algae. Corals also have a relationship with the fish that rest or take refuge within their branches, and which excrete nutrients that make the corals grow faster. The majority of corals form large colonies, although there are species that live singly. Their accumulated skeletons make up large coral reefs and atolls. The Great Barrier Reef, to the northeast of Australia, is about 1,600 km/1,000 mi long, has a total area of 20,000 sq km/7,700 sq mi, and adds 50 million metric tons of calcium to the reef each year. The world's reefs cover an estimated 620,000 sq km/240,000 sq mi.

Coral reefs provide a habitat for a diversity of living organisms. By 1997 some 93,000 species had been identified. One-third of the world's marine fishes live in reefs. The world's first global survey of coral reefs, carried out in 1997 found that around 95 percent of reefs had experienced some damage from overfishing, pollution, dynamiting, poisoning, and the dragging of ships' anchors.

Since the 1990s coral reefs have been destroyed by previously unknown diseases. The so-called white plague attacked 17 species of coral in the Florida Keys in 1995. The rapid wasting disease, discovered in 1997, affects coral reefs from Mexico to Trinidad. In the Caribbean, the fungus *Aspergillus* attacks sea fans, a soft coral. It was estimated in 1997 that around 90 percent of the coral around the Galàpagos Islands had been destroyed as a result of *bleaching,* a whitening of coral reefs that occurs when the colored algae evacuate the coral. This happens either because the corals produce toxins that are harmful to the algae or because they do not produce sufficient nutrients. Without the algae, the coral crumbles and dies away. Bleaching is widespread all over the Caribbean and the Indo-Pacific.

Fringing reefs are so called because they form close to the shores of continents or islands, with the living animals mainly occupying the outer edges of the reef. *Barrier reefs* are separated from the shore by a salt-water lagoon, which may be as much as 30 km/20 mi wide; there are usually navigable passes through the barrier into the lagoon. *Atolls* resemble a ring surrounding a lagoon, and do not enclose an island. They are usually formed by the gradual subsidence of an extinct volcano, with the coral growing up from where the edge of the island once lay.

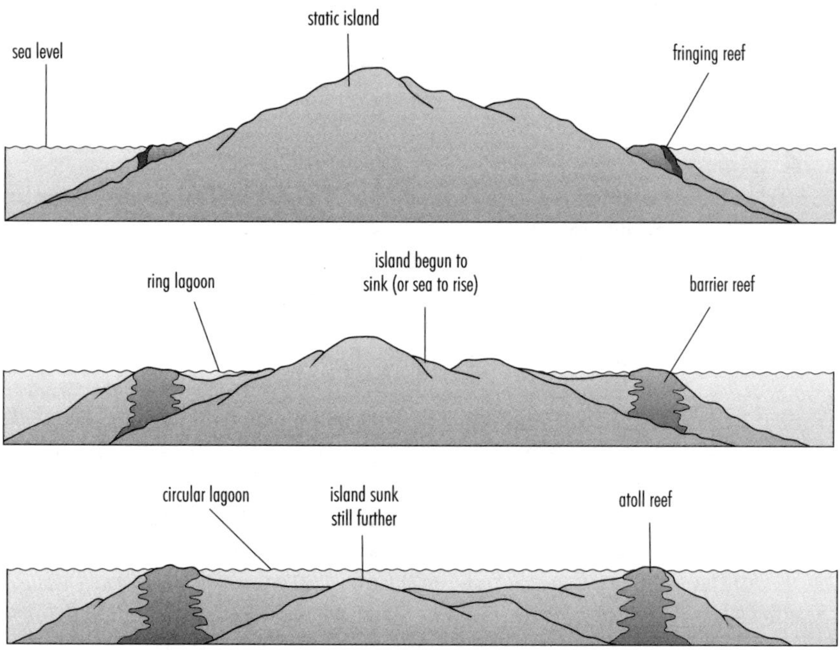

cordierite

A silicate mineral, $(Mg,Fe)_2Al_4Si_5O_{18}$, blue to purplish in color. It is characteristic of metamorphic rocks formed from clay sediments under conditions of low pressure but moderate temperature; it is the mineral that forms the spots in spotted slate and spotted hornfels.

cordillera

Spanish for "chain of mountains." A group of mountain ranges and their valleys, all running in a specific direction, formed by the continued convergence of two tectonic plates (*see* **plate tectonics**) along a line. The term is applied especially to the principal mountain region of a continent. The Andes of South America are an example.

core

The innermost part of Earth. It is divided into an outer core, which begins at a depth of 2,898 km/1,800 mi, and an inner core, which begins at a depth of 4,982 km/3,095 mi. Both parts are thought to consist of iron-nickel alloy. The outer core is liquid and the inner core is solid. The fact that seismic shear waves (*see* **seismic wave**) disappear at the mantle–outer core boundary indicates that the outer core is molten, because shear waves cannot travel through fluid. Scientists infer the iron-nickel rich composition of the core from Earth's density and its moment of inertia. The temperature of the core, as estimated from the melting point of iron at high pressure, is thought to be at least 4,000°C/7,232°F, but remains controversial. Earth's magnetic field is believed to be the result of the motions involving the inner and outer cores.

Coriolis effect

The effect of the earth's rotation on the atmosphere and on all objects on the earth's surface. In the Northern Hemisphere it causes moving objects and currents to be deflected to the right; in the Southern Hemisphere it causes deflection to the left. The effect is named for its discoverer, French mathematician Gaspard de Coriolis (1792–1843).

corrasion

The grinding away of solid rock surfaces by particles carried by water, ice, and wind. It is generally held to be the most significant form of erosion. As the eroding particles are carried along they become eroded themselves due to the process of attrition.

corrie

The Scottish term for a steep-sided hollow in the mountainside of a glaciated area. The Welsh call this a *cwm,* while in North America it is known as a *cirque.* The weight and movement of the ice has ground out the bottom and worn back the sides. A corrie is open at the front, and its sides and back are formed of arêtes. There may be a lake in the bottom, called a *tarn.*

A corrie is formed as follows: (1) snow accumulates in a hillside hollow (enlarging the hollow by nivation), and turns to ice; (2) the hollow is deepened by abrasion and plucking; (3) the ice in the corrie rotates under the influence of gravity, deepening the hollow still further; (4) because the ice is thinner and moves more slowly at the foot of the hollow, a rock lip forms; (5) when the ice melts, a lake or tarn may be formed in the corrie. The steep back wall may be severely weathered by freeze-thaw weathering, providing material for further abrasion.

corrosion

An alternative term for *solution,* the process by which water dissolves rocks such as limestone.

corundum

Native aluminum oxide, Al_2O_3, the hardest naturally occurring mineral known, apart from diamond (corundum rates 9 on the Mohs scale of hardness). Lack of cleavage also increases its durability. Corundum crystals are barrel-shaped prisms of the trigonal system. Varieties of gem-quality corundum are ruby (red) and sapphire (any color other than red, usually blue). Poorer-quality and synthetic corundum is used in industry, for example as an abrasive. Corundum forms in silica-poor igneous and metamorphic rocks. It is a constituent of *emery,* which is metamorphosed bauxite.

crag

In previously glaciated areas, a large lump of rock that a glacier has been unable to wear away. As the glacier passed up and over the crag, weaker rock on the far side was largely protected from erosion and formed a tapering ridge, or *tail,* of debris. An example of a crag-and-tail feature is found in Edinburgh in Scotland; Edinburgh Castle was built on the crag (Castle Rock), which dominates the city beneath.

crater

A bowl-shaped depression in the ground, usually round and with steep sides. Craters are formed by explosive events such as the eruption of a volcano, the explosion of a bomb, or the impact of a meteorite.

The moon has more than 300,000 craters over 1 km/6 mi in diameter, formed by meteorite bombardment; similar craters on Earth have mostly been worn away by erosion. Craters are found on many other bodies in the solar system.

Studies at the Jet Propulsion Laboratory in California have shown that craters produced by impact or by volcanic activity have distinctive shapes, enabling astronomers to distinguish likely methods of crater formation on planets in the solar system. Unlike volcanic craters, impact craters have a raised rim and central peak and are almost always circular, irrespective of the meteorite's angle of incidence.

craton

The core of a continent, a vast tract of highly deformed metamorphic rock around which the continent has been built; also known as the *shield.* Intense mountain-building periods shook these shield areas in Precambrian times before stable conditions set in.

Cratons exist in the hearts of all the continents, a typical example being the Canadian Shield.

Cretaceous

From the Latin *creta,* meaning *chalk,* the period of geological time approximately 144.2 million to 65 million years ago. It is the last period of the Mesozoic era, during which angiosperm (seed-bearing) plants evolved, and dinosaurs reached a peak before their extinction at the end of the period. The north European chalk, which forms the white cliffs of Dover, England, was deposited during the latter half of the Cretaceous period.

crevasse

A deep crack in the surface of a glacier; it can reach several meters in depth. Crevasses often occur where a glacier flows over the break of a slope, because the upper layers of ice are unable to stretch and cracks result. Crevasses may also form at the edges of glaciers, owing to friction with the bedrock.

crust

The outermost part of the structure of Earth, consisting of two distinct parts, the *oceanic crust* and the *continental crust.* The oceanic crust is on average about 10 km/6.2 mi thick and consists mostly of basaltic types of rock. By contrast, the continental crust is largely made of granite and is more complex in its structure. Because of the movements of plate tectonics, the oceanic crust is in no place older than about 200 million years. However, parts of the continental crust are over 3 billion years old.

Beneath a layer of surface sediment, the oceanic crust is made up of a layer of basalt, followed by a layer of gabbro. The composition of the oceanic crust overall shows a high proportion of *si*licon and *ma*gnesium oxides, hence named *sima* by geologists. The continental crust varies in thickness from about 40 km/25 mi to 70 km/45 mi, being deeper beneath mountain ranges. The surface layer consists of many kinds of sedimentary and igneous rocks. Beneath lies a zone of metamorphic rocks built on a thick layer of granodiorite. *Si*licon and *al*uminum oxides dominate the composition and the name *sial* is given to continental crustal material.

cryolite

A rare granular crystalline mineral (sodium aluminum fluoride), Na_3AlF_6, used in the electrolytic reduction of bauxite to aluminum. It is chiefly found in Greenland.

crystal system

All known crystalline substances crystallize in one of the seven crystal systems defined by symmetry. The elements of symmetry used for this purpose are: (1) planes of *mirror symmetry,* across which a mirror image is seen, and (2) axes of *rotational symmetry,* about which, in a 360° rotation of the crystal, equivalent faces are seen twice, three, four, or six times. To be assigned to a particular crystal system, a mineral must possess a certain minimum symmetry, but it may also possess additional symmetry elements. Because crystal symmetry is related to internal structure, a given mineral will always crystallize in the same system, although the crystals may not always grow into precisely the same shape. When two minerals have the same chemical composition but different internal structures (for example graphite and diamond, or quartz and cristobalite), they will generally have different crystal systems.

cuesta

An alternative name for *escarpment.*

current

The flow of a body of water or air, or of heat, moving in a definite direction. Ocean currents are fast-flowing currents of sea water generated by the wind or by variations in water density between two areas. They are partly responsible for transferring heat from the equator to the Poles and thereby evening out the global heat imbalance. There are three basic types of ocean current: *drift currents* are broad and slow moving; *stream currents* are narrow and swift moving; and *upwelling currents* bring cold, nutrient-rich water from the ocean bottom.

Stream currents include the Gulf Stream and the Japan (or Kuroshio) Current. Upwelling currents, such as the Gulf of Guinea Current and the Peru (Humboldt) Current, provide food for plankton, which in turn supports fish and sea birds. At approximate intervals of five to eight years, the Peru Current that runs from the Antarctic up the west coast of South America, turns warm, with heavy rain and rough seas, and has disastrous results (as in 1982–1983) for Peruvian wildlife and for the anchovy industry. The phenomenon is called El Niño, Spanish for "the boy child" with particular reference to the Christ Child, because it occurs toward Christmas.

cwm

See **corrie.**

cyclone

An alternative name for a *depression,* an area of low atmospheric pressure. A severe cyclone that forms in the tropics is called a tropical cyclone or hurricane. *See also* **atmospheric pressure.**

dating

The science of determining the age of geological structures, rocks, and fossils, and placing them in the context of geological time. The techniques are of two types: *relative dating* and *absolute dating.*

Relative dating can be carried out by identifying fossils of creatures that lived only at certain times (marker fossils), and by looking at the physical relationships of rocks to other rocks of a known age.

Absolute dating uses the process of radiometric dating to measure how much of a rock's radioactive elements has changed since the rock was formed.

Deep-Sea Drilling Project

A research project initiated by the United States in 1968 to sample the rocks of the ocean crust. In 1985 it became known as the Ocean Drilling Program (ODP).

deep-sea trench

Another term for ocean trench.

delta

The tract of land at a river's mouth, composed of silt deposited as the water slows on entering the sea. Familiar examples of large deltas are those of the Mississippi,

Ganges and Brahmaputra, Rhône, Po, Danube, and Nile; the shape of the Nile delta is like the Greek letter *delta* Δ, and thus gave rise to the name.

The *arcuate* (fan-shaped) *delta* of the Nile is only one form. Others are *birdfoot deltas,* like that of the Mississippi which is a seaward extension of the river's levee system; and *tidal deltas,* like that of the Mekong, in which most of the material is swept to one side by sea currents.

denudation

The natural loss of soil and rock debris, blown away by wind or washed away by running water, that lays bare the rock below. Over millions of years, denudation causes a general lowering of the landscape.

depression

Also *cyclone* or *low.* In meteorology, a region of low atmospheric pressure. In midlatitudes a depression forms as warm moist air from the tropics mixes with cold dry polar air, producing warm and cold boundaries (fronts) and unstable weather—low clouds and drizzle, showers, or fierce storms. The warm air, being less dense, rises above the cold air to produce the area of low pressure on the ground. Air spirals in toward the center of the depression in a counterclockwise direction in the Northern Hemisphere, clockwise in the Southern Hemisphere, generating winds up to gale force. Depressions tend to travel eastward and can remain active for several days.

A deep depression is one in which the pressure at the center is very much lower than its surroundings; it produces very strong winds, as opposed to a shallow depression, in which the winds are comparatively light. A severe depression in the tropics is called a hurricane, tropical cyclone, or typhoon, and is a great danger to shipping. A *tornado* is a very intense, rapidly swirling depression, with a diameter of only a few hundred meters or so.

a typical depression showing low pressure at the center

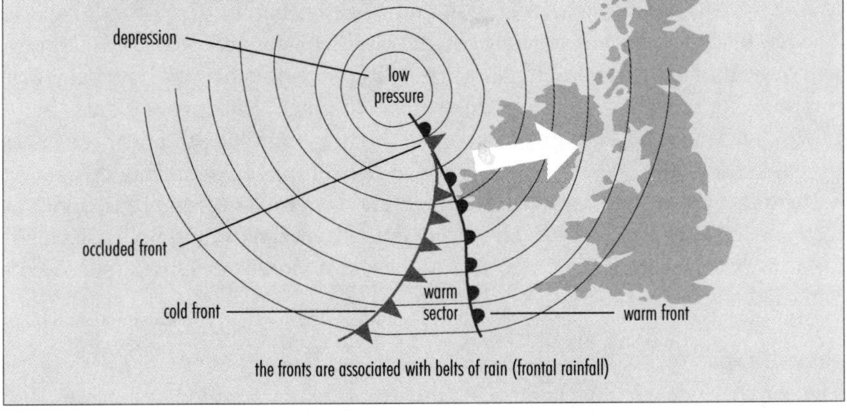

desalination

The removal of salt, usually from sea water, to produce fresh water for irrigation or drinking. Distillation has usually been the method adopted, but in the 1970s a cheaper process, using certain polymer materials that filter the molecules of salt from the water by reverse osmosis, was developed.

Desalination plants occur along the shores of the Middle East where fresh water is in short supply.

desert

An arid area with sparse vegetation (or, in rare cases, almost no vegetation). Soils are poor, and many deserts include areas of shifting sands. Deserts can be either hot or cold. Almost 33 percent of the earth's land surface is desert, and this proportion is increasing.

Tropical deserts, belts of latitudes from 5° to 30°, are caused by the descent of air that is heated over the warm land and therefore has lost its moisture. Other natural desert types are the *continental deserts,* such as the Gobi, that are too far from the sea to receive any moisture; *rain-shadow deserts,* such as California's Death Valley, that lie in the lee of mountain ranges, where the ascending air drops its rain only on the windward slopes; and *coastal deserts,* such as the Namib in southern Africa, where cold ocean currents cause local dry air masses to descend. Desert surfaces are usually rocky or gravelly, with only a small proportion being covered with sand. Deserts can be created by changes in climate, or by the human-aided process of desertification.

Characteristics common to all deserts include irregular rainfall of less than 250 mm/19.75 in per year, very high evaporation rates—often 20 times the annual precipitation, and low relative humidity and cloud cover. Temperatures are more variable; tropical deserts have a big diurnal temperature range and very high daytime temperatures (58°C/136.4°F has been recorded at Azizia in Libya), whereas midlatitude deserts have a wide annual range and much lower winter temperatures (in the Mongolian Desert the mean temperature is below the freezing point for half the year).

Desert soils are infertile, lacking in humus and generally gray or red in color. The few plants capable of surviving such conditions are widely spaced, scrubby, and often thorny. Long-rooted plants (phreatophytes) such as the date palm and mesquite commonly grow along dry stream channels. Salt-loving plants (halophytes) such as saltbushes grow in areas of highly saline soils and near the edges of playas (dry saline lakes). Others, such as the xerophytes are drought resistant and survive by remaining leafless during the dry season or by reducing water losses with small waxy leaves. They frequently have shallow and widely branching root systems and store water during the wet season (for example, succulents and cacti with pulpy stems).

desertification

The spread of deserts by changes in climate, or by human-aided processes. Desertification can sometimes be reversed by special planting (marram grass, trees) and by the use of water-absorbent plastic grains, which, added to the soil, enable crops to be grown.

Natural causes of desertification include decreased rainfall, increased temperatures, lowering of the water table, and soil erosion.

The human-aided processes leading to desertification include overgrazing, destruction of forest belts, and exhaustion of the soil by intensive cultivation without restoration of fertility—all of which may be prompted by the pressures of an expanding population or by concentration in land ownership. About 135 million people are directly affected by desertification, mainly in Africa, the Indian subcontinent, and South America. The Sahel region in Africa is one example.

destructive margin

Also *convergent margin.* In plate tectonics, the expression on Earth's surface where two plates of lithosphere converge. Usually one plate (the denser of the two) is forced to dive below the other, forming a subduction zone. The descending plate melts to form bodies of magma that may then rise to the surface through cracks and faults to form a series of volcanoes called a *volcanic arc.* If the two plates of lithosphere are capped by buoyant continental crust, subduction of the crust does not occur. Instead it crumples gradually to form mountains, such as the Himalayas.

devil wind

A minor form of tornado, usually occurring in fine weather, formed from rising thermals of warm air (as is a cyclone). A fire creates a similar updraft, allowing formation of a *fire devil* or *firestorm.* These may occur in oil refinery fires, or in the firebombings of cities, as happened in Dresden, Germany, in World War II.

Devonian

A period of geological time between 408 million and 360 million years ago, the fourth period of the Paleozoic era. Many desert sandstones from North America and Europe date from this time. The first land plants flourished in the Devonian period, corals were abundant in the seas, amphibians evolved from air-breathing fish, and insects developed on land. The name comes from the county of Devon in southwest England, where Devonian rocks were first studied.

dew

Precipitation in the form of moisture that collects on the ground. It forms after the temperature of the ground has fallen below the dew point of the air in contact with it. As the temperature falls during the night, the air and its water vapor become chilled, and condensation takes place on the cooled surfaces.

dew point

The temperature at which the air becomes saturated with water vapor. At temperatures below the dew point, the water vapor condenses out of the air as droplets. If the droplets are large they become deposited on the ground as dew; if small they remain in suspension in the air and form mist or fog.

diagenesis

The physical and chemical changes by which a sediment becomes a sedimentary rock. The main processes involved include compaction of the grains, and the cementing of the grains together by the growth of new minerals deposited by percolating groundwater.

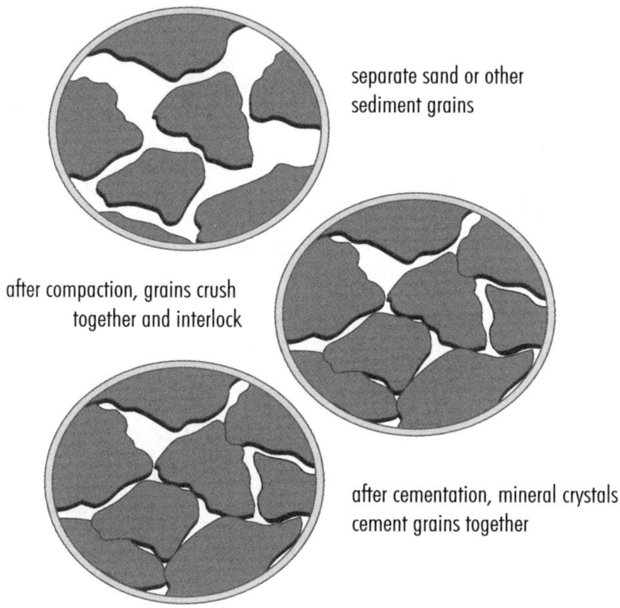

diamond

A generally colorless, transparent mineral, one of several crystalline forms of carbon, C. It is regarded as a precious gemstone, and is the hardest substance known (10 on the Mohs scale). Industrial diamonds, which may be natural or synthetic, are used for cutting, grinding, and polishing.

Diamond crystallizes in the cubic system as octahedral crystals, some with curved faces and striations. The high refractive index of 2.42 and the high dispersion of light, or "fire," account for the spectral displays seen in polished diamonds.

Sources. Diamonds may be found as alluvial diamonds on or close to the earth's surface in riverbeds or dried watercourses; on the sea bottom (off southwest Africa); or, more commonly, in diamond-bearing volcanic pipes composed of "blue ground," kimberlite or lamproite, where the original matrix has penetrated the earth's crust from great depths. They are sorted from the residue of crushed ground by x-ray and other recovery methods.

Varieties. There are four chief varieties of diamond: (1) well-crystallized transparent stones, colorless or only slightly tinted, valued as gems; (2) boart, poorly crystallized or inferior diamonds; (3) balas, an industrial variety, extremely hard and tough; and (4) carbonado, or industrial diamond, also called black diamond or carbon, which is opaque, black, or gray, and very tough. Industrial diamonds are also produced synthetically from graphite. Some synthetic diamonds conduct heat 50 percent more efficiently than natural diamonds and are five times greater in strength. This is a great advantage in their use to disperse heat in electronic and telecommunication devices and in the production of laser components.

Practical uses. Because diamonds act as perfectly transparent windows and do not absorb infrared radiation, they were used aboard NASA space probes to Venus in 1978. The tungsten-carbide tools used in steel mills are cut with industrial diamond tools.

Cutting. Rough diamonds are often dull or greasy before being polished; around 50 percent are considered "cuttable" (all or part of the diamond may be set into jewelry). Gem diamonds are valued by weight (carat), cut (highlighting the stone's optical properties), color, and clarity (on a scale from internally flawless to having a large inclusion clearly visible to the naked eye). They are sawn and polished using a mixture of oil and diamond powder. The two most popular cuts are the brilliant, for thicker stones, and the marquise, for shallower ones. India is the world's chief cutting center.

Noted rough diamonds include the Cullinan, or Star of Africa (3,106 carats, over 500 g/17.5 oz before cutting, South Africa, 1905); Excelsior (995.2 carats, South Africa, 1893); and Star of Sierra Leone (968.9 carats, Yengema, 1972).

Experiments. A moderate force applied to the small tips of two opposing diamonds can be used to attain extreme pressures of millions of atmospheres or more (*see* **diamond-anvil cell**), allowing scientists to subject small amounts of material to conditions that exist deep within planet interiors.

diamond-anvil cell

A device composed of two opposing cone-shaped diamonds that, when squeezed together by a lever arm, exert extreme pressures. The high pressures result from applying force to the small areas of the opposing diamond faces. The device is used to determine the properties of materials at pressures corresponding to those of planetary interiors. One discovery made with the diamond-anvil cell is $MgSiO_3$—perovskite, the predominant mineral of Earth's lower mantle.

diapirism

A geological process in which a particularly light rock, such as rock salt, punches upward through the heavier layers above. The resulting structure is called a salt dome, and oil is often trapped in the curled-up rocks at each side.

diatom

A microscopic alga found in all parts of the world in either fresh or marine waters. Diatoms consist of single cells that secrete a hard cell wall made of silica (class, Bacillariophyta). The cell wall of a diatom is made up of two overlapping valves known as frustules, which are impregnated with silica, and which fit together like the lid and body of a pillbox. Diatomaceous earths (diatomite) are made up of the valves of fossil diatoms, and are used in the manufacture of dynamite and in the rubber and plastics industries.

dike

A sheet of igneous rock created by the intrusion of magma (molten rock) across layers of preexisting rock. (By contrast, a *sill* is intruded between layers of rock.) It may form a ridge when exposed on the surface if it is more resistant than the rock into which it intruded. A dike is also a human-made embankment built along a coastline (for example, in the Netherlands) to prevent the flooding of lowland coastal regions.

diorite

An igneous rock intermediate in composition between mafic (consisting primarily of dark-colored minerals) and felsic (consisting primarily of light-colored minerals)—the coarse-grained plutonic equivalent of andesite. Constituent minerals include feldspar and amphibole or pyroxene with only minor amounts of quartz.

dip

The angle at which a structural surface, such as a fault or a bedding plane, is inclined from horizontal. Measured at right angles to the strike of that surface, it is used together with strike to describe the orientation of geological features in the field. Rocks that are dipping have usually been affected by folding. (*See also* **strike.**)

discharge

In a river, the volume of water passing a certain point per unit of time. It is usually expressed in cubic meters per second (cumecs). The discharge of a particular river channel may be calculated by multiplying the channel's cross-sectional area (in square meters) by the velocity of the water (in meters per second).

In the United Kingdom most small rivers have an average discharge of less than 1 cumec. The highest discharge usually occurs in early spring when melted snow combines with rainfall to increase the amount of water entering a river. Discharges are studied carefully by scientists as they can signal when a river is in danger of flooding. Channel efficiency is an important factor in determining the likelihood of flooding at times of high discharge.

distributary

A river that has branched away from a main river. Distributaries are most commonly found on a delta, where the very gentle gradient and large amounts of silt deposited encourage channels to split.

Channels are said to be *braided* if they branch away from a river and rejoin it at a later point. (*See* **braiding.**)

divergent margin

In plate tectonics, another term for constructive margin.

doldrums

An area of low atmospheric pressure along the equator, in the intertropical convergence zone where the northeast and southeast trade winds converge. The doldrums are characterized by calm or very light winds, during which there may be sudden squalls and stormy weather. For this reason the areas are avoided as far as possible by sailing ships.

dolomite

In mineralogy, a white mineral with a rhombohedral structure, calcium magnesium carbonate ($CaMg(CO_3)_2$). Dolomites are common in geological successions of all ages and are often formed when limestone is changed by the replacement of the mineral calcite with the mineral dolomite.

In sedimentology, dolomite is a type of limestone rock in which the calcite content is replaced by the mineral dolomite. Dolomite rock may be white, gray, brown, or reddish in color, commonly crystalline. It is used as a building material. The region of the Alps known as the Dolomites is a fine example of dolomite formation.

dome

A geologic feature that is the reverse of a basin. It consists of anticlinally folded rocks that dip in all directions from a central high point, resembling an inverted but usually irregular cup.

Such structural domes are the result of pressure acting upward from below to produce an uplifted portion of the crust. Domes are often formed by the upwelling of plastic materials such as salt or magma. The salt domes along the North American Gulf Coast were produced by upwelling ancient sea-salt deposits, while the Black Hills of South Dakota are the result of structural domes pushed up by intruding igneous masses.

downhole measurement

Any one of a number of experiments performed by instruments lowered down a borehole. Such instruments may be detectors that study vibrations passing through the rock from a generator, or may measure the electrical resistivity of the rock or the natural radiation.

Geophysicists measured heat flow at 10,000 boreholes worldwide from 1986 to 1994 to gather information on climate change. Surface air temperature affects the temperature of underlying rock. Heat travels so slowly that the first 500 m/ 1,640 ft of the earth's crust provides a record of the ground temperature for the last thousand years.

drainage basin

Also called a catchment area, the area of land drained by a river system (a river and its tributaries). It includes a the surface runoff in the hydrological cycle, as well as the water table. The edge of a drainage basin is called the *watershed.* A drainage basin is an example of an *open system* because it is open to inputs from outside, such as precipitation, and is responsible for outputs out of the system, such as output of water into the sea and evaporation of water into the atmosphere.

dune

A mound or ridge of wind-drifted sand common on coasts and in deserts. Loose sand is blown and bounced along by the wind, up the windward side of a dune. The sand particles then fall to rest on the lee side, while more are blown up from the windward side. In this way a dune moves gradually downwind.

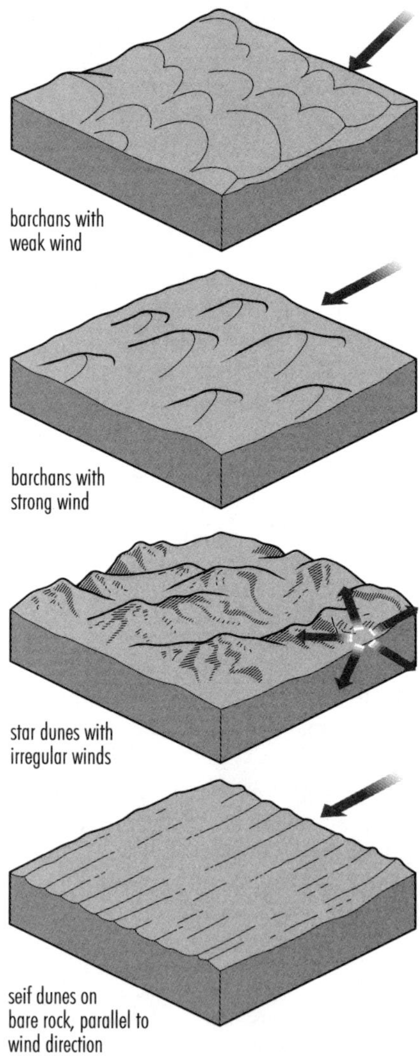

In sandy deserts, the typical crescent-shaped dune is called a *barchan. Seif dunes* are longitudinal and lie parallel to the wind direction, and *star-shaped dunes* are formed by irregular winds.

dust bowl

An area in the Great Plains region of North America (Texas to Kansas) that suffered extensive wind erosion as the result of drought and poor farming practice in once-fertile soil. Much of the topsoil was blown away in the droughts of the 1930s and the 1980s.

Similar dust bowls are being formed in many areas today, noticeably across Africa, because of overcropping and overgrazing.

dust devil

A small dust storm caused by intense local heating of the ground in desert areas, particularly the Sahara. The air swirls upward, carrying fine particles of dust with it.

Earth

The third planet from the sun. It is almost spherical, flattened slightly at the Poles, and is composed of three concentric layers: the core, the mantle, and the crust. About 70 percent of the surface (including the north and south polar icecaps) is covered with water. The Earth is surrounded by a life-supporting atmosphere and is the only planet on which life is known to exist.

Mean distance from the sun: 149,500,000 km/92,860,000 mi.
Equatorial diameter: 12,756 km/7,923 mi.
Circumference: 40,070 km/24,900 mi.
Rotation period: 23 hr 56 min 4.1 sec.

Year (complete orbit, or sidereal period): 365 days 5 hr 48 min 46 sec. Earth's average speed around the sun is 30 kps/18.5 mps; the plane of its orbit is inclined to its equatorial plane at an angle of 23.5°, the reason for the changing seasons.

Atmosphere: nitrogen (N_2) 78.09 percent; oxygen (O_2) 20.95 percent; argon (Ar) 0.93 percent; carbon dioxide (CO_2) 0.03 percent; and less than 0.0001 percent neon (Ne), helium (He), krypton (Kr), hydrogen (H_2), xenon (Xe), ozone (O_3), and radon.

Surface: land surface 150,000,000 sq km/57,500,000 sq mi (greatest height above sea level 8,872 m/29,118 ft Mount Everest); water surface 361,000,000 sq km/139,400,000 sq mi (greatest depth 11,034 m/36,201 ft Mariana Trench in the Pacific). The interior is thought to be an inner core about 2,600 km/1,600 mi in diameter, of solid iron and nickel; an outer core about 2,250 km/1,400 mi thick, of molten iron and nickel; and a mantle of mostly solid rock about 2,900 km/1,800 mi thick, separated from the earth's crust by the Mohorovicic discontinuity. The crust and the topmost layer of the mantle

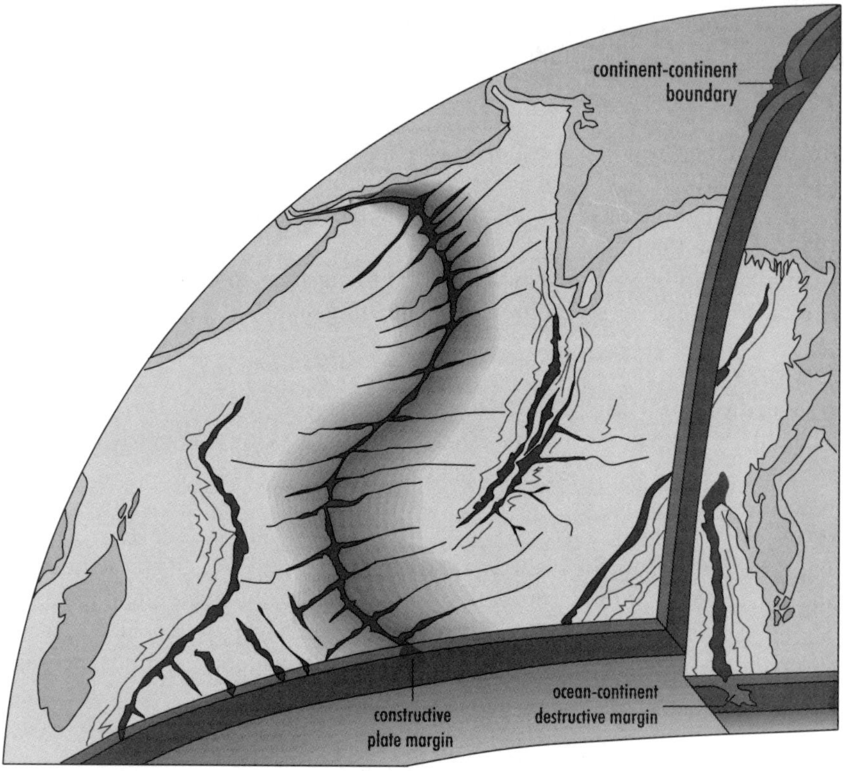

Earth: crust

form about 12 major moving plates, some of which carry the continents. The plates are in constant, slow motion, called *tectonic drift.* Geophysicists in the United States announced in 1996 that they had detected a difference in the spinning time of the earth's core and the rest of the planet; the core is spinning slightly faster.

Satellite: the moon

Age: 4.6 billion years. The earth was formed with the rest of the solar system by consolidation of interstellar dust. Life began 3.5 billion to 4 billion years ago.

earth science

The scientific study of the planet Earth as a whole. The mining and extraction of minerals and gems, the prediction of weather and earthquakes, the pollution of the atmosphere, land, and oceans, and the forces that shape the physical world all fall within its scope of study. The emergence of the discipline reflects scientists' concern that an understanding of the global aspects of the earth's structure and its past will hold the key to how humans affect its future, ensuring that its resources are used in a sustainable way. Earth science is a synthesis of several traditional subjects such as geology, meteorology, oceanography, geophysics, geochemistry, and paleontology.

earthquake

Abrupt motion that propagates through the earth and along its surfaces. Earthquakes are caused by the sudden release in rocks of strain accumulated over time as a result of tectonics. The study of earthquakes is called *seismology.* Most earthquakes occur along faults (fractures or breaks) and Benioff zones. Plate tectonic movements generate the major proportion: as two plates move past each other they can become jammed. When sufficient strain has accumulated, the rock breaks, releasing a series of elastic waves (seismic waves) as the plates spring free. The force of earthquakes (magnitude) is measured on the Richter scale, and their effect (intensity) on the Mercalli scale. The point at which an earthquake originates is the *seismic focus* or *hypocenter;* the point on the earth's surface directly above this is the *epicenter.*

Most earthquakes happen at sea and cause little damage. However, when severe earthquakes occur in highly populated areas they can cause great destruction and loss of life. The Alaskan earthquake of 27 March 1964 ranks as one of the greatest ever recorded, measuring 8.3 to 8.8 on the Richter scale. The San Andreas fault in California, where the North American and Pacific plates move past each other, is a notorious site of many large earthquakes. The 1906 San Francisco earthquake is among the most famous in history. Its magnitude was 8.3 on the Richter scale. The deadliest, most destructive earthquake in historical times is thought to have been in China in 1556. The New Madrid earthquake of 1811 was the strongest earthquake ever recorded in North America, although it occurred before the existence of the Richter scale. With its epicenter located in New Madrid, Missouri, the quake and its aftershocks could be felt as far away as Canada and affected an area of approximately 600,000 sq. km.

A reliable form of earthquake prediction has yet to be developed, although the seismic gap theory has had some success in identifying likely locations (*see* **seismic gap theory**). In 1987 a California earthquake was successfully predicted by measurement of underground pressure waves; prediction attempts have also involved the study of such phenomena as the change in gases issuing from the crust, the level of water in wells, slight deformation of the rock surface, a sequence of minor tremors, and the behavior of animals. The possibility of earthquake prevention is remote. However, rock slippage might be slowed at movement points, or promoted at stoppage points, by the extraction or injection of large quantities of water underground, because water serves as a lubricant. This would ease overall pressure.

Earthquakes have been responsible for moving the North Pole toward Japan at a rate of about 6 cm/2 in every 100 years. This is because most major earthquakes occur along the Pacific Rim, and tend to tilt the Pole toward their epicenters.

Ekman spiral effect

In oceanography, theoretical description of a consequence of the Coriolis effect on ocean currents, whereby currents flow at an angle to the winds that drive them.

It derives its name from the Swedish oceanographer Vagn Ekman (1874–1954). In the Northern Hemisphere, surface currents are deflected to the right of the wind direction. The surface current then drives the subsurface layer at an angle to its original deflection. Consequent subsurface layers are similarly affected, so that the effect decreases with increasing depth. The result is that most water is transported at about right angles to the wind direction. Directions are reversed in the Southern Hemisphere.

El Niño

A warm ocean surge of the Peru Current. Its Spanish name means "the (male) Child" and refers to the Christ child, because it tends to occur at Christmas. El Niño recurs about every five to eight years in the eastern Pacific off South America. It involves a change in the direction of ocean currents, which prevents the upwelling of cold nutrient-rich waters along the coast of Ecuador and Peru, killing fishes and plants. It is an important factor in global weather.

El Niño is believed to be caused by the failure of trade winds and, consequently, of the ocean currents normally driven by these winds. Warm surface waters then flow in from the east. The phenomenon can disrupt the climate of the area disastrously, and has played a part in causing famine in Indonesia, drought and bush fires in the Galápagos Islands, rainstorms in California and South America, and the destruction of Peru's anchovy harvest and wildlife in 1982–1983. El Niño contributed to algal blooms in Australia's drought-stricken rivers and an unprecedented number of typhoons in Japan in 1991. It is also thought to have caused the 1997 drought in Australia and contributed to certain ecological disasters such as bush fires in Indonesia.

El Niño usually lasts for about 18 months, but the 1990 occurrence lasted until June 1995; U.S. climatologists estimated this duration to be the longest in 2,000 years. The last prolonged El Niño of 1939–1941 caused extensive drought and famine in Bengal. It has been suggested that there might be a link between El Niño and global warming.

In a small way, El Niño affects the entire planet. The wind patterns of the 1998 El Niño have slowed the earth's rotation, adding 0.4 milliseconds to each day, an effect measured on the Very Long Baseline Interferometer (VLBI).

By examining animal fossil remains along the west coast of South America, U.S. researchers estimated in 1996 that El Niño began 5,000 years ago.

emerald

A clear green gemstone variety of the mineral beryl. It occurs naturally in Colombia, the Ural Mountains in Russia, Zimbabwe, and Australia. The green color is caused by the presence of the element chromium in the beryl.

emery

A black to grayish form of impure corundum that also contains the minerals magnetite and hematite. It is used as an abrasive. Emery occurs on the island of Naxos, Greece, and in Turkey.

environmental issues

Matters relating to the detrimental effects of human activity on the biosphere, their causes, and the search for possible solutions. Since the Industrial Revolution the demands made by both the industrialized and developing nations on the earth's natural resources are increasingly affecting the balance of the earth's resources. Over a period of time, some of these resources are renewable—trees can be replanted, soil nutrients can be replenished—but many resources, such as fossil fuels and minerals, are nonrenewable and in danger of eventual exhaustion. In addition, humans are creating many other problems that may endanger not only their own survival, but also that of other species. For instance, deforestation and air pollution are not only damaging and radically altering many natural environments, they also may be affecting the earth's climate by adding to the greenhouse effect and global warming, while water pollution is seriously affecting aquatic life, including fish populations, as well as human health.

Environmental pollution is normally taken to mean harm done to the natural environment by human activity. In fact, some environmental pollution can have natural sources, for example volcanic activity, which can cause major air pollution or water pollution and destroy flora and fauna. In terms of environmental issues, however, environmental pollution relates to human actions, especially in connection with energy resources. The demands of the industrialized nations for energy to power machines, provide light, heat, and so on are constantly increasing. The most versatile form of energy is electricity, which can be produced from a wide variety of other energy sources, such as the fossil fuels—coal, oil, and gas—and nuclear power produced from uranium. These are all nonrenewable resources and, in addition, their extraction, transportation, utilization, and waste products all give rise to pollutants of one form or another. The effects of these pollutants can have consequences not only for the local environment, but also at a global level.

Widespread effects of pollution. Many people think of air, water, and soil pollution as distinctly separate forms of pollution. However, each part of the global ecosystem—air, water, and soil—depends upon the others, and upon the plants and animals living within the environment. Thus, pollution that might appear to affect only one part of the environment is also likely to affect other parts. For example, the emission of vehicle exhausts or acid gases from a power plant might appear to harm only the surrounding atmosphere. But once released into the air they are carried by the prevailing winds, often for several hundred kilometers, before being deposited as acid rain. This can produce an enormous range of adverse effects across a very large area. For example: increased acidity levels in lakes and rivers are harmful to fish stocks and other aquatic life; physical damage to trees and other vegetation results in widespread destruction of forest areas; increased acidity of soils reduces the range of crops that can be grown, as well as decreasing production levels. Rocks such as limestone, both in the natural landscape and in buildings, are eroded—the effect of acid rain on some of the world's most important architectural structures is having disastrous consequences. In addition, acid rain in the form of aerosols (particles of liquid or solid suspended in a gas) or attached to smoke particles can cause respiratory problems in humans. Pollution of the Arctic atmosphere is creating Arctic haze—the result of aerosol emissions, such as dust, soot, and sulfate particles, originating in Europe.

Desertification. The destruction of fertile topsoil, and consequent soil erosion, as a result of human activity is becoming a worldwide problem. About 25 percent of the planet's land surface is now thought to be at risk due to increased demand from expanding populations. This damage and destruction results not only from increased demand for food, but also as a result of changes in agricultural practices. Desertification of vast areas, such as in the Sahel in northern Africa, have resulted from the replacement of traditional farming methods in these marginal lands for the present-day cultivation of cash crops such as groundnuts and cotton. The consequence has been that the soil has lost its fertility and the land has become arid. Similarly, changes in agricultural practices produced the dust bowl in the United States in the 1930s and, more recently, the move from mixed farming (cultivation of crops as well as livestock farming) to arable farming (cultivation of crops only) and the removal of hedges in order to enlarge fields for the use of modern agricultural machinery has resulted in the loss of topsoil in large areas of the English Fenlands.

Public awareness. The inspiration for the modern environmental movements came about from the publication in 1962 of Rachel Carson's book *Silent Spring,* in which she attacked the indiscriminate use of pesticides. This, combined with the increasing affluence of Western nations, which allowed people to look beyond their everyday needs, triggered an awareness of environmental issues on a global scale and resulted in the formation of the Green movement. In the mid-1960s, the detection of chlorofluorocarbons (CFCs) in the atmosphere by British scientist James Lovelock led to a realization of the damaging effects of ozone depletion and added to public concern for the environment, as did his development of the Gaia hypothesis, which views the earth as a single integrated and self-sustaining organism.

International measures. In 1972 the United Nations Environment Program (UNEP) was formed to coordinate international measures for monitoring and protecting the environment, and in 1985 the Vienna Convention for the Protection of the Ozone Layer, which promised international cooperation in research, monitoring, and the exchange of information on the problem of ozone depletion, was signed by 22 nations. Discussions arising out of this convention led to the signing in 1987 of the Montreal Protocol. In 1992 representatives of 178 nations met in Rio de Janeiro for the United Nations Conference on Environment and Development. Known as the "Earth Summit," this was one of the most important conferences ever held on environmental issues. United Nations members signed agreements on the prevention of global warming and the preservation of forests and endangered species, along with many other environmental issues. The United States did not sign or ratify some of these agreements.

Eocene

The second epoch of the Tertiary period of geological time, 56.5 million to 35.5 million years ago. Originally considered the earliest division of the Tertiary, the name means "early recent," referring to the early forms of mammals evolving at the time, following the extinction of the dinosaurs.

eolith

A naturally shaped or fractured stone found in Lower Pleistocene deposits and once believed by some scholars to be the oldest known human artifact type, dating to the pre-Paleolithic era. They are now recognized as not having been made by humans.

epoch

A subdivision of a geological period in the geological time scale. Epochs are sometimes given their own names (such as the Paleocene, Eocene, Oligocene, Miocene, and Pliocene epochs comprising the Tertiary period), or they are referred to as the late, early, or middle portions of a given period (as the Late Cretaceous or the Middle Triassic epoch).

Geological time is broken up into *geochronological units* of which the epoch is just one level of division. The hierarchy of geochronological divisions is eon, era, period, epoch, age, and chron. Epochs are subdivisions of periods and ages are subdivisions of epochs. Rocks representing an epoch of geological time comprise a *series.*

equator

Also called the *terrestrial equator,* the great circle whose plane is perpendicular to the earth's axis (the line joining the Poles). Its length is 40,092 km/24,901.8 mi, divided into 360 degrees of longitude. The equator encircles the broadest part of the earth, and represents 0° latitude. It divides the earth into two halves, called the *Northern* and the *Southern Hemispheres.*

The *celestial equator* is the circle in which the plane of the earth's equator intersects the celestial sphere.

era

Any of the major divisions of geological time, each including several periods, but smaller than an eon. The currently recognized eras all fall within the Phanerozoic eon—or the vast span of time, starting about 570 million years ago, when fossils are found to become abundant. The eras in ascending order are the Paleozoic, Mesozoic, and Cenozoic. We are living in the Recent epoch of the Quaternary period of the Cenozoic era.

Geological time is broken up into *geochronological units* of which the era is just one level of division. The hierarchy of geochronological divisions is eon, era, period, epoch, age, and chron. Eras are subdivisions of eons and periods are subdivisions of eras. Rocks representing an era of geological time comprise an *erathem.*

erosion

Wearing away of the earth's surface, caused by the breakdown and transportation of particles of rock or soil. (By contrast, *weathering* does not involve transportation.) Agents of erosion include the sea, rivers, glaciers, and wind. Water, consisting of sea waves and currents, rivers, and rain; ice, in the form of glaciers; and wind, hurling sand fragments against exposed rocks and moving dunes along, are the most potent forces of erosion. People also contribute to erosion by bad farming practices and the cutting down of forests, which can lead to the formation of dust bowls. There are several processes of erosion, including hydraulic action, corrasion, attrition, and solution.

erratic

A displaced rock that has been transported by a glacier or some other natural force to a site of different geological composition.

escarpment

Also *cuesta.* A large ridge created by the erosion of dipping sedimentary rocks. An escarpment has one steep side (scarp) and one gently sloping side (dip). Escarpments are common features of chalk landscapes. Certain features are associated with chalk escarpments, including dry valleys (formed on the dip slope), combes (steep-sided valleys on the scarp slope), and springs.

esker

A narrow, steep-walled ridge, often sinuous and sometimes branching, formed beneath a glacier. It is made of sands and gravels, and represents the course of a subglacial river channel. Eskers vary in height from 3–30 m/10–100 ft and can be up to 160 km/100 mi or so in length.

estuary

A river mouth widening into the sea, where fresh water mixes with salt water and tidal effects are felt.

Etesian wind

A north-northwesterly wind that blows from June to September in the eastern Mediterranean and Aegean Seas.

eustatic change

The worldwide rise or fall in sea level caused by a change in the amount of water in the oceans (by contrast, isostasy involves a rising or sinking of the land). During the last ice age, sea level fell because water became "locked-up" in the form of ice and snow, and less water reached the oceans.

evaporite

A sedimentary deposit precipitated on evaporation of salt water. With a progressive evaporation of seawater, the most common salts are deposited in a definite sequence: calcite (calcium carbonate), gypsum (hydrous calcium sulfate), halite (sodium chloride), and finally salts of potassium and magnesium.

Calcite precipitates when seawater is reduced to half its original volume, gypsum precipitates when the seawater body is reduced to one-fifth, and halite when the volume is reduced to one-tenth. Thus the natural occurrence of chemically precipitated calcium carbonate is common, of gypsum fairly common, and of halite less common.

Because of the concentrations of different dissolved salts in sea water, halite accounts for about 95 percent of the chlorides precipitated if evaporation is complete. More unusual evaporite minerals include borates (for example borax, hydrous sodium borate) and sulfates (for example glauberite, a combined sulfate of sodium and calcium).

exosphere

The uppermost layer of the atmosphere. It is an ill-defined zone above the thermosphere, beginning at about 700 km/435 mi and fading off into the vacuum of space. The gases are extremely thin, with hydrogen as the main constituent.

extrusive rock

Also *volcanic rock.* Igneous rock formed on the surface of the earth by volcanic activity. The term includes fine-grained crystalline or glassy rocks formed from hot lava quenched at or near Earth's surface and rocks composed of solid debris, called *pyroclastics,* deposited by explosive eruptions.

Large amounts of extrusive rock called *basalt* form at the earth's ocean ridges from lava that fills the void formed when two tectonic plates spread apart. Explosive volcanoes that deposit pyroclastics generally occur where one tectonic plate descends beneath another. andesite is often formed by explosive volcanoes. Magmas that give rise to pyroclastic extrusive rocks are explosive because they are viscous. The island of Montserrat, West Indies, is an example of an explosive volcano that spews pyroclastics of andesite composition. Magmas that produce crystalline or glassy volcanic rocks upon cooling are less viscous. The low viscosity allows the extruding lava to flow easily. Fluidlike lavas that flow from the volcanoes of the Hawaiian Islands have low viscosity and cool to form basalt.

facies

A body of rock strata possessing unifying characteristics indicative of the environment in which the rocks were formed. The term is also used to describe the environment of formation itself or unifying features of the rocks that comprise the facies.

Features that define a facies can include collections of fossils, sequences of rock layers, or the occurrence of specific minerals. Sedimentary rocks deposited at the same time, but representing different facies belong to a single *chronostratigraphic unit* (*see* **stratigraphy**). But these same rocks may belong to different *lithostratigraphic units.* For example, beach sand is deposited at the same time that mud is deposited further offshore. The beach sand eventually turns to sandstone while the mud turns to shale. The resulting sandstone and shale strata comprise two different facies, one representing the beach environment and the other the offshore environment, formed at the same time; the sandstone and shale belong to the same chronostratigraphic unit but distinct lithostratigraphic units.

The set of characteristics that distinguish one facies from another in a given chronostratigraphic unit is used to interpret local variations in environments that existed at the same time. Changes in environment can be deduced from a variety of features. One facies in a body of rock might consist of porous limestone containing fossil reef-building organisms in their living positions. This facies might laterally pass into a reef-flank facies of steeply dipping deposits of rubble from the reef, which in turn might grade into an interreef basin composed of fine, clayey limestone.

Ancient floods and migrations of the seashore up or down can also be traced by changes in facies.

fault

A fracture in the earth either side of which rocks have moved past one another. Faults involve displacements, or offsets, ranging from the microscopic scale to hundreds of kilometers/miles. Large offsets along a fault are the result of the accumulation of smaller movements (of meters/feet or less) over long periods of time. Large motions cause detectable earthquakes.

Faults are planar features. Fault orientation is described by the inclination of the fault plane with respect to horizontal (*see* **dip**) and its direction in the horizontal plane (*see* **strike**). Faults at high angle with respect to horizontal (in which the fault plane is steep) are classified as either *normal faults,* where one block has apparently moved downhill along the inclined fault plane, or *reverse faults,* where one block appears to have moved uphill along the fault plane. Normal faults occur where rocks on either side have moved apart. Reverse faults occur where rocks on either side have been forced together. A reverse fault that forms a low angle with the horizontal plane is called a *thrust fault.*

A *lateral fault,* or *tear fault,* occurs where the relative movement along the fault plane is sideways. A particular kind of fault found only in ocean ridges is the

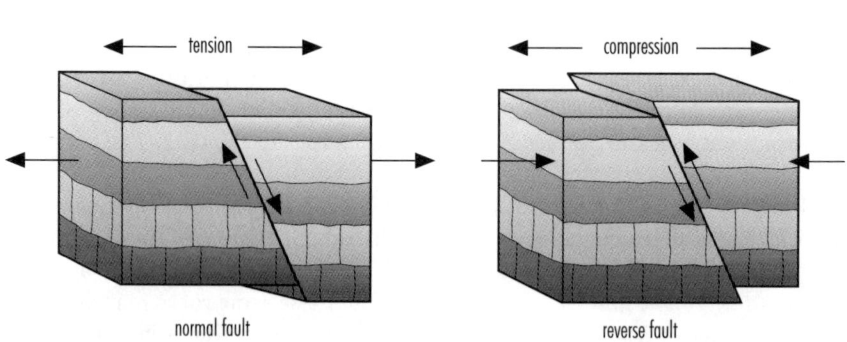

transform fault (a term coined by Canadian geophysicist John Tuzo Wilson in 1965). On a map, an ocean ridge has a stepped appearance. The ridge crest is broken into sections, each section offset from the next. Between each section of the ridge crest the newly generated plates are moving past one another, forming a transform fault.

Faults produce lines of weakness on the earth's surface (along their strike) that are often exploited by processes of weathering and erosion. Coastal caves and geos (narrow inlets) often form along faults and, on a larger scale, rivers may follow the line of a fault.

fault gouge

A soft, uncemented, pulverized claylike material found between the walls of a fault. It is created by grinding action (cataclasis) of the fault and subsequent chemical alteration of tiny mineral grains brought about by fluids that flow along the fault.

feldspar

A group of silicate minerals. Feldspars are the most abundant mineral type in the earth's crust. They are the chief constituents of igneous rock and are present in most metamorphic and sedimentary rocks. All feldspars contain silicon, aluminum, and oxygen, linked together to form a framework. Spaces within this framework structure are occupied by sodium, potassium, calcium, or occasionally barium, in various proportions. Feldspars form white, gray, or pink crystals and rank 6 on the Mohs scale of hardness.

The four extreme compositions of feldspar are represented by the minerals orthoclase, $KAlSi_3O_8$; albite, $NaAlSi_3O_8$; anorthite, $CaAl_2Si_2O_8$; and celsian, $BaAl_2Si_2O_8$. *Plagioclase feldspars* contain variable amounts of sodium (as in albite) and calcium (as in anorthite) with a negligible potassium content. *Alkali feldspars* (including orthoclase) have a high potassium content, less sodium, and little calcium.

The type of feldspar known as *moonstone* has a pearl-like effect and is used in jewelry. Approximately 4,000 metric tons of feldspar are used in the ceramics industry annually.

feldspathoid

Any of a group of silicate minerals resembling feldspars but containing less silica. Examples are nepheline ($NaAlSiO_4$ with a little potassium) and leucite ($KAlSi_2O_6$). Feldspathoids occur in igneous rocks that have relatively high proportions of sodium and potassium. Such rocks may also contain alkali feldspar, but they do not generally contain quartz because any free silica would have combined with the feldspathoid to produce more feldspar instead.

felsic rock

A plutonic rock composed chiefly of light-colored minerals, such as quartz, feldspar, and mica. It is derived from feldspar, lenad (meaning feldspathoid), and silica.

The term *felsic* also applies to light-colored minerals as a group, especially quartz, feldspar, and feldspathoids.

fetch

The distance of open water over which wind can blow to create waves. The greater the fetch, the more potential power waves have when they hit the coast. In the south and west of England the fetch stretches for several thousand kilometers, all the way to South America. This combines with the southwesterly prevailing winds to cause powerful waves and serious coastal erosion along south- and west-facing coastlines.

fiord

An alternative spelling of *fjord.*

fire clay

A clay with refractory characteristics (resistant to high temperatures), and hence suitable for lining furnaces (firebrick). Its chemical composition consists of a high percentage of silicon and aluminum oxides, and a low percentage of the oxides of sodium, potassium, iron, and calcium.

firn

Also *névé.* Snow that has lain on the ground for a full calendar year. Firn is common at the tops of mountains; for example, the Alps in Europe. After many years, compaction turns firn into ice and a glacier may form.

fjord

Also *fiord.* A narrow sea inlet enclosed by high cliffs. Fjords are found in Norway, New Zealand, and western parts of Scotland. They are formed when an overdeepened U-shaped glacial valley is drowned by a rise in sea level. At the mouth of the fjord there is a characteristic lip causing a shallowing of the water. This is due to reduced glacial erosion and the deposition of moraine at this point.

Fiordland is the deeply indented southwest coast of South Island, New Zealand; one of the most beautiful inlets is Milford Sound.

flash flood

A flood of water in a normally arid area brought on by a sudden downpour of rain. Flash floods are rare and usually occur in mountainous areas. They may travel many miles from the site of the rainfall.

Because of the suddenness of flash floods, little warning can be given of their occurrence. In 1972 a flash flood at Rapid City, South Dakota, killed 238 people along Rapid Creek.

flint

A compact, hard brittle mineral (a variety of chert), brown, black, or gray in color, found as nodules in limestone or shale deposits. It consists of cryptocrystalline (grains too small to be visible even under a light microscope) silica, SiO_2, principally in the crystalline form of quartz. Implements fashioned from flint were widely used in prehistory.

The best flint, used for Neolithic tools, is floorstone, a shiny black flint that occurs deep within chalk.

Because of their hardness (7 on the Mohs scale), flint splinters are used for abrasive purposes and, when ground into powder, added to clay during pottery manufacture. Flints have been used for making fire by striking the flint against steel, which produces a spark, and for discharging guns. Flints in cigarette lighters are made from cerium alloy.

flocculation

In soils, the artificially induced coupling together of particles to improve aeration and drainage. Clay soils, which have very tiny particles and are difficult to work, are often treated in this way. The method involves adding more lime to the soil.

flood plain

The area of periodic flooding along the course of river valleys. When river discharge exceeds the capacity of the channel, water rises over the channel banks and floods the adjacent low-lying lands. As water spills out of the channel some alluvium (silty material) will be deposited on the banks to form levees (raised river banks). This water will slowly seep into the flood plain, depositing a new layer of rich fertile alluvium as it does so.

Many important flood plains, such as the inner Niger delta in Mali, occur in arid areas where their exceptional productivity has great importance for the local economy.

A flood plain (sometimes called an *inner delta*) can be regarded as part of a river's natural domain, statistically certain to be claimed by the river at repeated intervals. By plotting floods that have occurred and extrapolating from these data we can speak of 10-year floods, 100-year floods, 500-year floods, and so forth, based on the statistical probability of flooding across certain parts of the flood plain.

Even the most energetic flood-control plans (such as dams, dredging, and channel modification) will sometimes fail, and using flood plains as the site of towns and villages is always laden with risk. It is more judicious to use flood plains in ways compatible with flooding, such as for agriculture or parks. Flood plain features include meanders and oxbow lakes.

flooding

The inundation of land that is not normally covered with water. Flooding from rivers commonly takes place after heavy rainfall or in the spring after winter snows have melted. The river's discharge (volume of water carried in a given period) becomes too great, and water spills over the banks onto the surrounding flood plain. Small floods may happen once a year. These are called *annual floods* and are said to have a 1-year return period. Much larger floods may occur on average only once every 50 years.

Flooding is least likely to occur in an efficient channel that is semicircular in shape (*see* **channel efficiency**). Flooding can also occur at the coast in stormy conditions (*see* **storm surge**) or when there is an exceptionally high tide.

fluidization

Making a mass of solid particles act as a fluid by agitation or by gas passing through. Much earthquake damage is attributed to fluidization of surface soils during the earthquake shock. Another example is ash flow during volcanic eruption.

fluorite

Also *fluorspar.* A glassy, brittle halide mineral, calcium fluoride, CaF_2, forming cubes and octahedra; colorless when pure, otherwise violet, blue, yellow, brown, or green. Fluorite is used as a flux in iron and steel making; colorless fluorite is used in the manufacture of microscope lenses. It is also used for the glaze on pottery, and as a source of fluorine in the manufacture of hydrofluoric acid.

fluorspar

Another name for the mineral fluorite.

fluvial

Of or pertaining to streams or rivers. A *fluvial deposit* is sedimentary material laid down by a stream or river, such as a sandstone or conglomerate (coarse-grained clastic sedimentary rock composed of rounded pebbles of preexisting rock cemented in a fine-grained sand or clay matrix).

fluvioglacial

Of a process or landform, associated with glacial meltwater. Meltwater, flowing beneath or ahead of a glacier, is capable of transporting rocky material and creating a variety of landscape features, including eskers, kames, and outwash plains.

flux

In smelting, a substance that combines with the unwanted components of the ore to produce a fusible slag, which can be separated from the molten metal. For example, the mineral fluorite, CaF_2, is used as a flux in iron smelting; it has a low melting point and will form a fusible mixture with substances of higher melting point such as silicates and oxides.

In soldering, a flux is a substance that improves the bonding properties of solder by removing contamination from metal surfaces and preventing their oxidation, and by reducing the surface tension of the molten solder alloy. For example, with solder made of lead-tin alloys, the flux may be resin, borax, or zinc chloride.

focus

The point within the earth's crust at which an earthquake originates. The point on the surface that is immediately above the focus is called the epicenter.

fog

Clouds that collect at the surface of the earth, composed of water vapor that has condensed on particles of dust in the atmosphere. Clouds and fog are both caused by the air temperature falling below the dew point. The thickness of fog depends on the number of water particles it contains. Officially, fog refers to a condition when visibility is reduced to 1 km/0.6 mi or less, and mist or haze to that giving a visibility of 1–2 km or about 1 mi.

There are two types of fog. An *advection fog* is formed by the meeting of two currents of air, one cooler than the other, or by warm air flowing over a cold surface. Sea fogs commonly occur where warm and cold currents meet and the air above them mixes. A *radiation fog* forms on clear, calm nights when the land surface loses heat rapidly (by radiation); the air above is cooled to below its dew point and condensation takes place. A mist is produced by condensed water particles, and a haze by smoke or dust.

In drought areas, for example, Baja California, Canary Islands, Cape Verde Islands, Namib Desert, Peru, and Chile, coastal fogs enable plant and animal life to survive without rain and are a potential source of water for human use (by means of water collectors exploiting the effect of condensation). Industrial areas uncontrolled by pollution laws have a continual haze of smoke over them, and if the temperature falls suddenly, a dense yellow smog forms. At some airports since 1975 it has been possible for certain aircraft to land and take off blind in fog, using radar navigation.

föhn

Also *foehn.* A warm dry wind that blows down the leeward slopes of mountains. The air heats up as it descends because of the increase in pressure, and it is dry because all the moisture was dropped on the windward side of the mountain. A similar wind, chinook, is found on the eastern slopes of the Rocky Mountains in North America.

fold

A bend in beds or layers of rock. If the bend is arched up in the middle it is called an *anticline;* if it sags downward in the middle it is called a *syncline.* The line along which a bed of rock folds is called its *axis.* The axial plane is the plane joining the axes of successive beds.

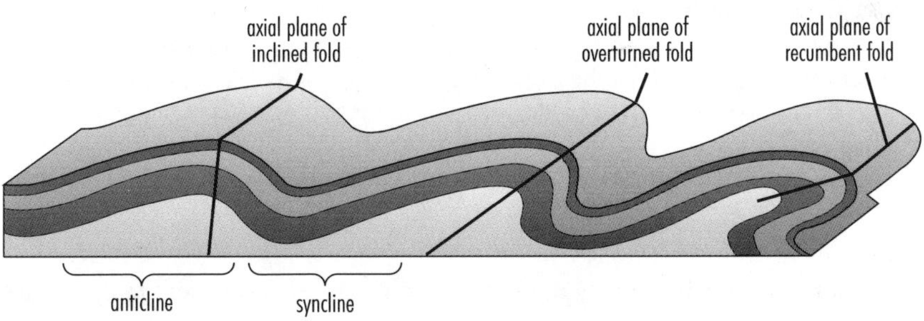

foredeep

An elongated structural basin lying inland from an active mountain system and receiving sediment from the rising mountains. According to plate tectonic theory, a mountain chain forming behind a subduction zone along a continental margin develops a foredeep or gently sloping trough parallel to it on the landward side. Foredeeps form rapidly and are usually so deep initially that the sea floods them through gaps in the mountain range. As the mountain system evolves, sediments choke the foredeep, pushing out marine water. As marine sedimentation stops, only nonmarine deposits from the rapidly eroding mountains are formed. These consist of alluvial fans and also rivers, flood plains, and related environments inland.

Before the advent of plate tectonic theory, such foredeep deposits and changes in sediments had been interpreted as sedimentary troughs, called geosynclines, that were supposed ultimately to build upward into mountains.

fossil

Latin *fossilis,* meaning *dug up.* A cast, impression, or the actual remains of an animal or plant preserved in rock. Fossils were created during periods of rock formation, caused by the gradual accumulation of sediment over millions of years at the bottom of the seabed or an inland lake. Fossils may include footprints, an internal cast, or external impression. A few fossils are preserved intact, as with mammoths fossilized in Siberian ice, or insects trapped in tree resin that is today amber. The study of fossils is called paleontology. Paleontologists are able to deduce much of the geological history of a region from fossil remains.

About 250,000 fossil species have been discovered—a figure that is believed to represent less than 1 in 20,000 of the species that ever lived. *Microfossils* are so small they can only be seen with a microscope. They include the fossils of pollen, bone fragments, bacteria, and the remains of microscopic marine animals and plants, such as foraminifera and diatoms.

freeze-thaw

A form of physical weathering, common in mountains and glacial environments, caused by the expansion of water as it freezes. Water in a crack freezes and expands in volume by 9 percent as it turns to ice. This expansion exerts great pressure on the rock, causing the crack to enlarge. After many cycles of freeze-thaw, rock fragments may break off to form scree slopes.

For freeze-thaw to operate effectively, the temperature must fluctuate regularly above and below 0°C/32°F. It is therefore uncommon in areas of extreme and perpetual cold, such as the polar regions.

fringing reef

A coral reef that is attached to the coast without an intervening lagoon.

front

In meteorology, the boundary between two air masses of different temperature or humidity. A *cold front* marks the line of advance of a cold air mass from below, as it displaces a warm air mass; a *warm front* marks the advance of a warm air mass as it rises up over a cold one. Frontal systems define the weather of the midlatitudes, where warm tropical air is constantly meeting cold air from the Poles.

Warm air, being lighter, tends to rise above the cold; its moisture is carried upward and usually falls as rain or snow, hence the changeable weather conditions at fronts. Fronts are rarely stable and move with the air mass. An *occluded front* is a composite form and occurs where a cold front catches up with a warm front and merges with it.

frost

A condition of the weather that occurs when the air temperature is below freezing, 0°C/32°F. Water in the atmosphere is deposited as ice crystals on the ground or exposed objects. As cold air is heavier than warm, ground frost is more common than hoarfrost, which is formed by the condensation of water particles in the same way that dew collects.

frost hollow

A depression or steep-sided valley in which cold air collects on calm, clear nights. Under clear skies, heat is lost rapidly from ground surfaces, causing the air above to cool and flow downhill (as katabatic wind) to collect in valley bottoms. Fog may form under these conditions and, in winter, temperatures may be low enough to cause frost.

frost shattering

See **freeze-thaw.**

fuller's earth

A soft, greenish-gray rock resembling clay, but without clay's plasticity. It is formed largely of clay minerals, rich in montmorillonite, but a great deal of silica is also present. Its absorbent properties make it suitable for removing oil and grease, and it was formerly used for cleaning fleeces, known as *fulling.* It is still used in the textile industry, but its chief application is in the purification of oils. Beds of fuller's earth are found in the southern United States, Germany, Japan, and the United Kingdom. Beds of fuller's earth in the United Kingdom are linked to volcanic activity in Mesozoic times, when fine volcanic ash settled in water.

gabbro

A mafic (consisting primarily of dark-colored crystals) igneous rock formed deep in the earth's crust. It contains pyroxene and calcium-rich feldspar, and may contain small amounts of olivine and amphibole. Its coarse crystals of dull minerals give it a speckled appearance.

Gabbro is the plutonic version of basalt (that is, derived from magma that has solidified below the earth's surface), and forms in large, slow-cooling intrusions.

galena

A mineral consisting of lead sulfide, PbS, the chief ore of lead. It is lead gray in color, has a high metallic luster and breaks into cubes because of its perfect cubic cleavage. It may contain up to 1 percent silver, and so the ore is sometimes mined for both metals. Galena occurs mainly among limestone deposits in Australia, Mexico, Russia, Kazakhstan, the United Kingdom, and the United States.

gangue

That part of an ore deposit that is not itself economically valuable; for example, calcite may occur as a gangue mineral with galena.

garnet

A group of silicate minerals with the formula $X_3Y_3(SiO_4)_3$, where X is calcium, magnesium, iron, or manganese, and Y is usually aluminum or sometimes iron or chromium. Garnets are used as semiprecious gems (usually pink to deep red) and as abrasives. They occur in metamorphic rocks such as gneiss and schist.

Garnets consisting of neodymium, yttrium, and aluminum (referred to as *Nd-YAG*) produce infrared laser light. Nd-YAG lasers are inexpensive and used widely in industry and scientific research.

gastrolith

A stone that was once part of the digestive system of a dinosaur or other extinct animal. Rock fragments were swallowed to assist in the grinding process in the dinosaur digestive tract, much as some birds now swallow grit and pebbles to grind food in their crop. Once the animal has decayed, smooth round stones remain—often the only clue to their past use is the fact that they are geologically different from their surrounding strata.

gelifluction

A type of solifluction (downhill movement of water-saturated topsoil) associated with frozen ground.

gem

Any mineral valuable by virtue of its durability (hardness), rarity, and beauty, cut and polished for ornamental use, or engraved. Of 120 minerals known to have been used as gemstones, only about 25 are in common use in jewelry today; of these, the diamond, emerald, ruby, and sapphire are classified as precious, and all the others are semiprecious; for example, the topaz, amethyst, opal, and aquamarine.

Among the synthetic precious stones to have been successfully produced are rubies, sapphires, emeralds, and diamonds (first produced by the General Electric company in the United States in 1955). Pearls are not technically gems.

With the exception of the diamond, most stones are valued for their color. However, this is often due to the presence of an impurity, and not a property of the mineral itself. For example, rubies are formed when the mineral corundum contains small amounts of chromium, whereas sapphires are formed when small amounts of iron and titanium oxides are present. The most common impurities are probably compounds of iron, manganese, and copper. Exposure to light makes some gems change or lose their color altogether; certain types of turquoise and topaz are particularly liable to do this.

geochemistry

The science of chemistry as it applies to geology. It deals with the relative and absolute abundances of the chemical elements and their isotopes in the earth, and also with the chemical changes that accompany geologic processes.

geochronology

The branch of geology that deals with the dating of the earth by studying its rocks and contained fossils. The geological time chart is a result of these studies. Absolute dating methods involve the measurement of radioactive decay over time in certain chemical elements found in rocks, whereas relative dating methods establish the sequence of deposition of various rock layers by identifying and comparing their contained fossils.

geode

A subspherical cavity into which crystals have grown from the outer wall into the center. Geodes often contain very well-formed crystals of quartz (including amethyst), calcite, or other minerals.

geodesy

Methods of surveying the earth for making maps and correlating geological, gravitational, and magnetic measurements. Geodesic surveys, formerly carried out by means of various measuring techniques on the surface, are now commonly made by using radio signals and laser beams from orbiting satellites (*see* **global positioning system**).

geography

The study of the earth's surface; its topography, climate, and physical conditions, and how these factors affect people and society. It is usually divided into *physical geography,* dealing with landforms and climates, and *human geography,* dealing with the distribution and activities of peoples on Earth.

geological time

A time scale embracing the history of the earth from its physical origin to the present day. Geological time is traditionally divided into eons (archean or Archeozoic, Proterozoic, and Phanerozoic in ascending chronological order), which in turn are subdivided into eras, periods, epochs, ages, and finally chrons.

The terms *eon, era, period, epoch, age,* and *chron* are *geochronological units* representing intervals of geological time. Rocks representing an interval of geological time comprise a *chronostratigraphic unit.* Each of the hierarchical geochronological terms has a chronostratigraphic equivalent. Thus, rocks formed during an eon (a geochronological unit) are members of an *eonothem* (the chronostratigraphic unit equivalent of eon). Rocks of an era belong to an *erathem*. The chronostratigraphic equivalents of period, epoch, age, and chron are *system, series, stage,* and *chronozone,* respectively.

geology

The science of the earth—its origin, composition, structure, and history. It is divided into several branches: *mineralogy* (the minerals of Earth), *petrology* (rocks), *stratigraphy* (the deposition of successive beds of sedimentary rocks), *paleontology* (fossils), and *tectonics* (the deformation and movement of the earth's crust).

Geology is regarded as part of earth science, a more widely embracing subject that brings in meteorology, oceanography, geophysics, and geochemistry.

geomorphology

A branch of geology, developed in the late nineteenth century, dealing with the morphology, or form, of the earth's surface; nowadays it is also considered to be an integral part of physical geography. Geomorphological studies involve investigating the nature and origin of surface landforms, such as mountains, valleys, plains, and plateaus, and the processes that influence them. These processes include the effects of tectonic forces, weathering, running water, waves, glacial ice, and wind, which result in the erosion, transportation, and deposition of rocks and soils. The underlying dynamics of these forces are the energy derived from the earth's gravitational field, the flow of solar energy through the hydrological cycle, and the flow of heat from the earth's molten interior. The mechanisms of these processes are both destructive and constructive; out of the destruction or modification of one landform another will be created. In addition to the natural processes that mold landforms, human activity can produce changes, either directly or indirectly, and cause the erosion, transportation, and deposition of rocks and soils, for example, by poor land-management practices and techniques in farming and forestry, and in the mining and construction industries.

Geomorphology deals with changes in landforms over seconds or eons, and in spatial scales ranging from undulations to mountains. For example, the formation of mountain ranges takes place over millions of years, as the earth's crust cools and solidifies and the resulting layers, or plates, are folded, uplifted, or deformed by the seismic activity of the underlying magma (*see* **plate tectonics**). The gouging out of river valleys by glacial erosion is a gradual process that takes place over millennia. On the other hand, volcanic eruptions, by the ejection of rocks and gases and the rapid flow of molten lava down a mountainside, create instantaneous changes to landforms, as with the recent volcanic eruptions on the island of Montserrat in the West Indies. Similarly, the eruption of undersea volcanoes can result in the sudden birth of islands, while the consequent and rapidly moving tidal waves, tsunamis, can produce the unexpected inundation and destruction of low-lying coastal regions in their path.

Erosion and weathering. Landforms are changed by the processes of weathering, erosion, transportation, and deposition. For example, the shape of a mountain range is largely the result of erosive processes that progressively remove material from the range. These include weathering and soil-forming processes, such as the physical, chemical, or organic breakdown of rocks into small pieces by the action of wind, rain, temperature changes, plants, and bacteria. The form that erosion takes, and the degree to which it takes place, can vary with rock type. Unconsolidated sands and gravel are more easily eroded than solid granite, while rocks such as limestone are worn down by chemical processes, rather than by physical forces. The weathering process does not involve particle transportation, except under the effects of gravity.

Transportation and deposition. After materials have been broken down and loosened by weathering, they are transported by mass movement, wind action, or running water, and deposited to new locations. Glaciers transport materials embedded in them, winds lift dust particles and carry them over great distances, precipitation falling on sloping land shifts soils downhill, water currents carry materials along a riverbed or out to sea. Through deposition, the particles accumulate elsewhere; rivers and glaciers carve valleys and deposit eroded material in plains and deltas, desert winds wear away rock and form huge sand dunes, waves erode rocky shorelines and create sandy beaches. River deltas such as those of the Nile and the Ganges, which are formed by the buildup of silt at the point where the river meets the sea, demonstrate the cumulative effects on landform of transportation and deposition.

Concepts and subdisciplines. At the end of the nineteenth century, the American geologist William Morris Davis advanced a unifying concept called the *geomorphic cycle.* He believed that landforms progressed from a youthful stage of high rugged mountains to a more mature stage of rounded forms, eventually to be worn down in old age to almost level plains. This theory does not hold up under current insights, and any specific landscape can be regarded only as the balance between whatever forces of uplift and erosion are operating at a given time. Thus, the current nature of a region does not allow us to reconstruct its past or predict its future; mountain building (orogeny) is a long, drawn-out, intermittent, and uneven process, and even if the progression from youth to old age were an uninterrupted sequence, this progression could produce different landforms, depending on climatic variables. The study of the different processes and variables involved has given rise to a number of subdisciplines such as evolutionary geomorphology, climatic geomorphology, structural geomorphology, tectonic geomorphology, process geomorphology, and applied geomorphology.

Applications. Geomorphology has many practical applications. An understanding of landforms and the processes involved in their formation and development provides a basis for the mapping of soils and the location of resources such as minerals and fossil fuels. It can also help to minimize the environmental damage caused by mining and poor farming practices. Land clearance, combined with harmful tillage techniques, can result in soil erosion, which in turn can rob fertile land of its productive topsoil and lead to desertification, as in the creation of the dust bowl in North America in the 1930s. The use of windbreaks and correct planting and cultivation techniques can prevent this erosion, increase agricultural production, and at the same time help to conserve the existing landform. Past projects for the control of coastal erosion and flooding have sometimes led to an intensification of the problems rather than their alleviation. A greater knowledge of the mechanics of offshore waves and currents can lead to a better understanding of their long-term effects, and eventually to the development of more effective methods of control. The aim of much current research in geomorphology is to increase our understanding of the complexity and interrelatedness of environmental factors. This, in turn, will aid the protection of existing landforms, and increase our ability to predict the occurrences of natural hazards such as avalanches, river flooding, and coastal erosion, and assist in the control of their consequences.

geophysics

The branch of earth science that uses physics to study the earth's surface, interior, and the atmosphere and hydrosphere. Specialties include seismology, paleomagnetics, and tectonophysics.

geyser

A natural spring that intermittently discharges an explosive column of steam and hot water into the air due to the buildup of steam in underground chambers. One of the most remarkable geysers is Old Faithful, in Yellowstone National Park, Wyoming. Geysers also occur in New Zealand and Iceland.

glacial deposition

The laying down of rocky material once carried by a glacier and the product of glacial erosion. When ice melts, it deposits the material that it has been carrying. The material deposited by a glacier is called *till,* or in Britain *boulder clay.* It comprises angular particles of all sizes from boulders to clay that are unsorted and lacking in stratification.

Unstratified till can be molded by ice to form *drumlins,* egg-shaped hills. At the snout of the glacier, till piles up to form a ridge called a *terminal moraine.* Small depositional landforms may also result from glacial deposition, such as *kames* (small mounds) and *kettle holes* (small depressions, often filled with water).

Stratified till that has been deposited by meltwater is termed *fluvioglacial,* because it is essentially deposited by running water. Meltwater flowing away from a glacier carries some of the till many miles away. This sediment becomes rounded (by the water) and, when deposited, forms a gently sloping area called an *outwash plain.* Several landforms owe their existence to meltwater (fluvioglacial landforms) and include the long ridges called *eskers,* which form parallel to the direction the ice flows. Meltwater may fill depressions eroded by the ice to form *ribbon lakes.*

Glacial deposits occur in many different locations beneath the ice (subglacial), inside it (englacial), on top of it (supraglacial), at the side of it (marginal), and in front of it (proglacial). In periglacial environments on the margins of an

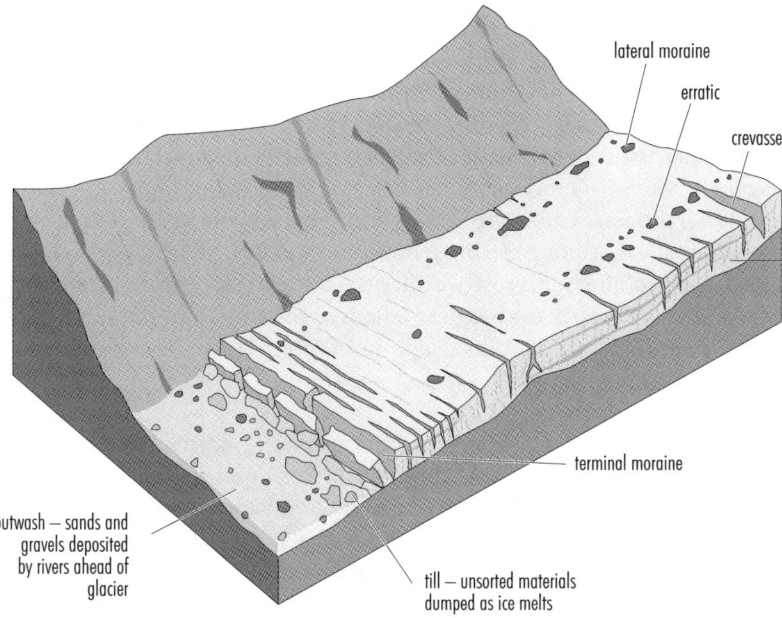

glacial deposition

ice sheet, frost weathering (the alternate freezing and thawing of ice in cracks in the rock) etches the outlines of rock outcrops, exploiting joints and areas of weakness, and results in aprons of scree.

glacial erosion

The wearing down and removal of rocks and soil by a glacier. Glacial erosion forms impressive landscape features, including glacial troughs (U-shaped valleys), arêtes (steep ridges), corries (enlarged hollows), and pyramidal peaks (high mountain peaks with concave faces).

Erosional landforms result from abrasion and plucking of the underlying bedrock. Abrasion is caused by the lodging of rock debris in the sole of the glacier, followed by friction and wearing away of the bedrock as the ice moves. The action is similar to that of sandpaper attached to a block of wood. The results include the polishing and scratching of rock surfaces to form powdered rock flour, and subparallel scratches or striations which indicate the direction of ice movement. Plucking is a form of glacial erosion restricted to the lifting and removal of blocks of bedrock already loosened by freeze-thaw activity in joint fracture.

The most extensive period of recent glacial erosion was the Pleistocene epoch in the Quaternary period when, over 2 million to 3 million years, the polar ice caps repeatedly advanced and retreated. More ancient glacial episodes are also preserved in the geological record, the earliest being in the middle Precambrian and the most extensive in Permo-Carboniferous times.

Larger landforms caused by glacial erosion generally possess a streamlined form, as in roche moutonnée and cirques. A common feature of glacial erosion is the glacial trough. Hanging valleys, with their tributary waterfalls, indicate extensive glacial erosion. The amount of lowering accomplished by glacial erosion has been estimated at 0.05–2.8 mm/0.002–0.11 in per year, a rate of lowering 10 to 20 times that associated with the action of rivers. Depositional landforms cover 10 percent of the earth's surface.

Periglacial processes result from frost and snow activity in areas on the margins of an ice sheet. Among the most important periglacial processes are frost-weathering and solifluction. Frost-weathering (the alternate freezing and thawing of ice in cracks in the rock) etches the outlines of rock outcrops, exploiting joints and areas of weakness, and results in aprons of scree.

glacial trough

Also *U-shaped valley.* A steep-sided, flat-bottomed valley formed by a glacier. The erosive action of the glacier and of the debris carried by it results in the formation not only of the trough itself but also of a number of associated features, such as truncated spurs (projections of rock that have been sheared off by the ice) and hanging valleys (smaller glacial valleys that enter the trough at a higher level than the trough floor). Features characteristic of glacial deposition, such as drumlins and eskers, are commonly found on the floor of the trough, together with linear lakes called *ribbon lakes.*

glacier

A tongue of ice, originating in mountains in snowfields above the snowline, which moves slowly downhill and is constantly replenished from its source. The geographic features produced by the erosive action of glaciers (glacial erosion) are characteristic and include glacial troughs (U-shaped valleys), corries, and arêtes. In lowlands, the laying down of rocky debris carried by glaciers (glacial deposition) produces a variety of landscape features, such as moraines, eskers, and drumlins.

Glaciers form where annual snowfall exceeds annual melting and drainage (*see* **glacier budget**). The area at the top of the glacier is called the *zone of accumulation.* The lower area of the glacier is called the *ablation zone.* In the zone of accumulation, the snow compacts to ice under the weight of the layers above and moves downhill under the force of gravity. The ice moves plastically under pressure, changing its shape and crystalline structure permanently. Partial melting of ice at the sole of the glacier also produces a sliding component of glacial movement, as the ice travels over the bedrock. In the ablation zone, melting occurs and glacial till is deposited.

When a glacier moves over an uneven surface, deep crevasses are formed in rigid upper layers of the ice mass; if it reaches the sea or a lake, it breaks up to form icebergs. A glacier that is formed by one or several valley glaciers at the base of a mountain is called a *piedmont glacier.* A body of ice that covers a large land surface or continent, for example Greenland or Antarctica, and flows outward in all directions is called an *ice sheet.*

In October 1996 a volcano erupted under Europe's largest glacier, Vatnajökull in Iceland, causing flooding.

glacier budget

In a glacier, the balance between accumulation (the addition of snow and ice to the glacier) and ablation (the loss of snow and ice by melting and evaporation). If accumulation exceeds ablation, the glacier will advance; if ablation exceeds accumulation, it will probably retreat.

The rate of advance and retreat of a glacier is usually only a few centimeters a year.

glaucophane

A blue amphibole, $Na_2(Mg,Fe,Al)_5Si_8O_{22}(OH)_2$. It occurs in glaucophane schists (blue schists), formed from the ocean floor basalt under metamorphic conditions of high pressure and low temperature; these conditions are believed to exist in subduction systems associated with destructive plate boundaries (*see* **plate tectonics**), and so the occurrence of glaucophane schists can indicate the location of such boundaries in geological history.

global positioning system (GPS)

A U.S. satellite-based navigation system, a network of 24 satellites in six orbits, each circling the earth once every 24 hours. Each satellite sends out a continuous time signal and an identifying signal. Position is determined from three intersecting measures of distance from the satellites to the user's receiver. Distance is determined by the transit time of the signal the satellite sends. To fix position, a user needs to be within range of four satellites, one to provide a reference signal to compute time accurately and the others to provide the three independent distances. The user's receiver calculates the position from the difference in time between receiving the signals from each satellite.

The position of the receiver can be calculated to better than 0.5 m/1.6 ft, although only the U.S. military can tap the full potential of the system. Other users can obtain a position to within 100 m/330 ft. This is accurate enough to be of use to boats, walkers, and motorists, and inexpensive, suitable receivers are on the market. GPS is also used in numerous scientific experiments as a means for determining position or time with accuracy. It is used in geophysics to monitor the movements of land, for example on either side of a fault, and in geodesy to determine elevation, for example the height of Mount Everest.

global warming

An increase in average global temperature of approximately 1°F/0.5°C over the past century. Global temperature has been highly variable in Earth's history and many fluctuations in global temperature have occurred in historical times, but this most recent episode of warming coincides with the spread of industrialization, prompting the hypothesis that it is the result of an accelerated greenhouse effect caused by atmospheric pollutants, especially carbon dioxide gas. Recent melting and collapse of the Larsen Ice Shelf, Antarctica, is a consequence of global warming. Melting of ice is expected to raise sea level in the coming decades.

Natural, perhaps chaotic, climatic variations have not been ruled out as the cause of the current global rise in temperature, but scientists are still assessing the likely influence of anthropogenic (human-made) pollutants. Assessing the impact of humankind on global climate is complicated by the natural variability on both geological and human time scales. The present episode of global warming has thus far still left England approximately 1°C cooler than during the peak of the so-called Medieval Warm Period (A.D. 1000–1400) The latter was part of a purely natural climatic fluctuation on a global scale. With respect to historical times, the interval between the Medieval Warm Period and the rise in temperatures we see today was unusually cold throughout the world.

In addition to a rise in average global temperature, global warming has caused seasonal variations to be more pronounced in recent decades. Examples are the most severe winter on record in the eastern United States in 1976–1977 and the record heat waves in the Netherlands and Denmark the following year.

A 1995 United Nations summit in Berlin agreed to take action to reduce gas emissions harmful to the environment. Delegates at the summit, from more than 120 countries, approved a two-year negotiating process aimed at setting specific targets and timetables for reducing nations' emissions of carbon dioxide and other greenhouse gases after the year 2000. The Kyoto Protocol of 1997 commits the world's industrialized countries to cut their annual emissions of harmful gases by 5.2 percent by 2012. As of end April 1998, 34 countries had ratified the Kyoto Protocol. These included all countries of the European Union, Canada, Norway, Brazil, and Japan. For the treaty to come into effect, 55 countries must ratify it. The 55 must include developed countries in sufficient numbers to account for 55 percent of global carbon emissions. The United States pledged to reduce its emissions to 7 percent less than its 1990 emissions, but this still has not been ratified by the U.S. Congress.

In June 1996 the Intergovernmental Panel on Climate Change—the official body of more than 2,000 top scientists set up by the world's governments—reported that global warming was already taking place and that human activities were probably to blame. Under the 1992 Convention on Climate Change, the industrialized countries pledged to level off emissions of carbon dioxide, the main cause of global warming, so that by the year 2000 they would be no higher than in 1990. However, evidence collected by the European Union, the World Energy Council, the International Energy Agency, and other international institutions showed that only three countries were likely to achieve this. Among Western countries, only Britain, Germany, and Luxembourg were expected to meet the target. In Britain, this achievement was largely due to the collapse of the coal industry and subsequent increase in the use of gas, a less polluting fuel. The International

Energy Agency estimated in 1996 that by the year 2000 emissions will be 17 percent higher than in 1990; by 2010 they will have risen by 49 percent. In 1997 the concentration of emissions reached 364 parts per million.

gneiss

A coarse-grained metamorphic rock, formed under conditions of high temperature and pressure, and often occurring in association with schists and granites. It has a foliated, or layered, structure consisting of thin bands of micas and/or amphiboles, dark in color alternating with bands of granular quartz and feldspar that are light in color. Gneisses are formed during regional metamorphism; *paragneisses* are derived from metamorphism of sedimentary rocks and *orthogneisses* from metamorphism of granite or similar igneous rocks.

gold rush

A large influx of gold prospectors to an area where gold deposits have recently been discovered. The result is a dramatic increase in population. Cities such as Johannesburg, Melbourne, and San Francisco either originated or were considerably enlarged by gold rushes.

Gondwanaland

Also *Gondwana.* A southern landmass formed 200 million years ago by the splitting of the single world continent, Pangaea. (The northern landmass was Laurasia.) It later fragmented into the continents of South America, Africa, Australia, and Antarctica, which then drifted slowly to their present positions. The baobab tree found in both Africa and Australia is a relic of this ancient land mass.

A database of the entire geology of Gondwanaland has been constructed by geologists in South Africa. The database, known as Gondwana Geoscientific Indexing Database (GO-GEOID), displays information as a map of Gondwana 155 million years ago, before the continents drifted apart.

gorge

A narrow steep-sided valley (or canyon) that may or may not have a river at the bottom. A gorge may be formed as a waterfall retreats upstream, eroding away the rock at the base of a river valley; or it may be caused by rejuvenation, when a river begins to cut downward into its channel once again (for example, in response to a fall in sea level). Gorges are common in limestone country, where they may be formed by the collapse of the roofs of underground caverns.

graded bedding

A sedimentary feature in which the sedimentary layer shows a gradual change in particle size, usually coarse at the bottom to fine at the top. It is useful for determining which way was up at the time the bed was deposited.

Grand Banks

The continental shelf in the North Atlantic off southeastern Newfoundland, where the shallow waters are rich fisheries, especially for cod.

granite

A coarse-grained intrusive igneous rock, typically consisting of the minerals quartz, feldspar, and biotite mica. It may be pink or gray, depending on the composition of the feldspar. Granites are chiefly used as building materials.

Granites often form large intrusions in the core of mountain ranges, and they are usually surrounded by zones of metamorphic rock (rock that has been altered by heat or pressure). Granite areas have characteristic moorland scenery. In exposed areas the bedrock may be weathered along joints and cracks to produce a *tor,* consisting of rounded blocks that appear to have been stacked upon one another.

graphite

A blackish gray, laminar crystalline form of carbon. It is used as a lubricant and as the active component of pencil lead. Graphite, like diamond and fullerene, is an allotrope of carbon. The carbon atoms are strongly bonded together in sheets, but the bonds between the sheets are weak, allowing other atoms to enter regions between the layers and cause them to slide over one another. Graphite has a very high melting point (3,500°C/6,332°F), and is a good conductor of heat and electricity. It absorbs neutrons and is therefore used to moderate the chain reaction in nuclear reactors.

gravel

A coarse sediment consisting of pebbles or small fragments of rock, originating in the beds of lakes and streams or on beaches. Gravel is quarried for use in road building, railroad ballast, and for an aggregate in concrete. It is obtained from quarries known as *gravel pits,* where it is often found mixed with sand or clay. Some gravel deposits also contain placer deposits of metal ores (chiefly tin) or free metals (such as gold and silver).

gravimetry

The study of the earth's gravitational field. Small variations in the gravitational field (gravimetric anomalies) can be caused by varying densities of rocks and structure beneath the surface. Such variations are measured by a device called a *gravimeter,* which consists of a weighted spring that is pulled further downward where the gravity is stronger (at a Bouguer anomaly). Gravimetry is used by geologists to map the subsurface features of the earth's crust, such as underground masses of heavy rock, like granite, or light rock such as salt.

Great Artesian Basin

The largest area of artesian water in the world. It underlies much of Queensland, New South Wales, and South Australia, and in prehistoric times formed a sea. It has an area of 1,750,000 sq km/676,250 sq mi.

great circle

A circle drawn on a sphere such that the diameter of the circle is a diameter of the sphere. On the earth, all meridians of longitude are half great circles; among the parallels of latitude, only the equator is a great circle.

The shortest route between two points on the earth's surface is along the arc of a great circle. These are used extensively as air routes, although on maps—owing to the distortion brought about by projection—they do not appear as straight lines.

greenhouse effect

A phenomenon of the earth's atmosphere by which solar radiation, trapped by the earth and reemitted from the surface as infrared radiation, is prevented from escaping by various gases in the air. Greenhouse gases trap heat because they readily absorb infrared radiation. The result is a rise in the earth's temperature (global warming). The main greenhouse gases are carbon dioxide, methane, and chlorofluorocarbons (CFCs) as well as water vapor. Fossil fuel consumption and forest fires are the principal causes of carbon dioxide buildup; methane is a by-product of agriculture (rice, cattle, sheep).

The United Nations Environment Program estimates that by 2025 average world temperatures will have risen by 1.5°C/2.7°F with a consequent rise of 20 cm/7.9 in in sea level. Low-lying areas and entire countries would be threatened by flooding and crops would be affected by the change in climate. However, predictions about global warming and its possible climatic effects are tentative and often conflict with each other.

At the 1992 Earth Summit it was agreed that by the year 2000 countries would stabilize carbon dioxide emissions at 1990 levels, but to halt the acceleration of

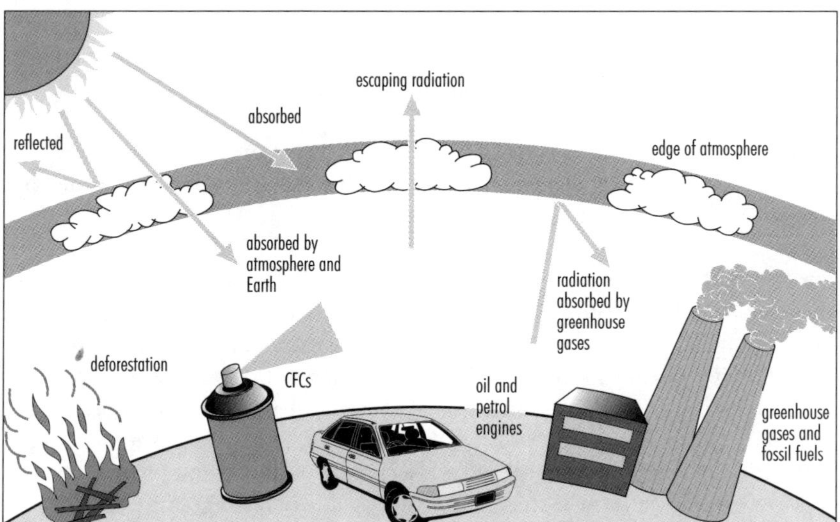

global warming, emissions would probably need to be cut by 60 percent. Any increases in carbon dioxide emissions are expected to come from transportation. The Berlin Mandate, agreed unanimously at the climate conference in Berlin in 1995, committed industrial nations to the continuing reduction of greenhouse gas emissions after 2000, when the existing pact to stabilize emissions runs out. The stabilization of carbon dioxide emissions at 1990 levels by 2000 will not be achieved by a number of developed countries, including Spain, Australia, and the United States, according to 1997 estimates. Australia is in favor of different targets for different nations, and refused to sign a communiqué at the South Pacific Forum meeting in the Cook Islands in 1997, which insisted on legally binding reductions in greenhouse gas emissions.

The concentration of carbon dioxide in the atmosphere is estimated to have risen by 25 percent since the Industrial Revolution, and 10 percent since 1950; the rate of increase is now 0.5 percent a year. Chlorofluorocarbon levels are rising by 5 percent a year, and nitrous oxide levels by 0.4 percent a year, resulting in a global warming effect of 0.5 percent since 1900, and a rise of about 0.1°C/3°F a year in the temperature of the world's oceans during the 1980s. Arctic ice was 6–7 m/20–23 ft thick in 1976 and had reduced to 4–5 m/13–17 ft by 1987.

A medium-sized power station, generating around 500 megawatts, produces 500 metric tons of carbon in the form of carbon dioxide every hour. It is possible to remove the gas from flue emissions and store it under pressure, although the energy needed to do so would amount to a fifth of the plant's electricity. In addition, the waste is bulky and must be disposed of safely.

It is also thought that aircraft vapor trails could be contributing to global warming. Aircraft release exhaust gases into the cold upper air, which freeze and form cirrus clouds. These clouds trap heat, increasing the warming of the atmosphere beneath them. German research indicates that one-tenth of the cirrus cloud over central Europe is produced by aircraft. While there is much concern over the amount of carbon dioxide released into the atmosphere by aircraft, cement kilns contribute more to the world's output of this gas than aircraft.

grike

An enlarged joint that separates blocks of limestone (clints) in a limestone pavement.

groundwater

Water collected underground in porous rock strata and soils; it emerges at the surface as springs and streams. The groundwater's upper level is called the *water table.* Sandy or other kinds of beds that are filled with groundwater are called *aquifers.* Recent estimates are that usable groundwater amounts to more than 90 percent of all the fresh water on Earth; however, keeping such supplies free of pollutants entering the recharge areas is a critical environmental concern.

Most groundwater near the surface moves slowly through the ground while the water table stays in the same place. The depth of the water table reflects the balance between the rate of infiltration, called *recharge,* and the rate of discharge at springs or rivers or pumped water wells. The force of gravity makes underground water run "downhill" underground just as it does above the surface. The greater the slope and the permeability, the greater the speed. Velocities vary from 100 cm/40 in per day to 0.5 cm/0.2 in.

groyne

A wooden or concrete barrier built at right angles to a beach in order to block the movement of material along the beach by longshore drift. Groynes are usually successful in protecting individual beaches, but because they prevent beach material from passing along the coast they can mean that other beaches, starved of sand and gravel, are in danger of being eroded away by the waves.

gulf

Any large sea inlet.

Gulf Stream

A warm ocean current that flows north from the warm waters of the Gulf of Mexico along the east coast of America, from which it is separated by a channel of cold water originating in the southerly Labrador current. Off Newfoundland, part of the current is diverted east across the Atlantic, where it is known as the North Atlantic Drift, dividing to flow north and south, and warming what would otherwise be a colder climate in the British Isles and northwest Europe.

At its beginning the Gulf Stream is 80–150 km/50–93 mi wide and up to 850 m/2,788 ft deep, and moves with an average velocity of 130 km/80 mi a day. Its temperature is about 26°C. As it flows northward, the current cools and becomes broader and less rapid.

guyot

A truncated or flat-topped seamount. Such undersea mountains are found throughout the abyssal plains of major ocean basins, and most of them are covered by an appreciable depth of water. They are believed to have started as volcanic cones formed near midoceanic ridges, in relatively shallow water, and to have been truncated by wave action as their tops emerged above the surface. They were then transported to deeper waters as the ocean floor moved away on either side of the ridge.

gypsum

A common sulfate mineral, composed of hydrous calcium sulfate, $CaSO_4.2H_2O$. It ranks 2 on the Mohs scale of hardness. Gypsum is used for making casts and molds, and for blackboard chalk. A fine-grained gypsum, called *alabaster,* is used for ornamental work. Burned gypsum is known as *plaster of Paris,* because for a long time it was obtained from the gypsum quarries of the Montmartre district of Paris.

gyre

A circular surface rotation of ocean water in each major sea (a type of current). Gyres are large and permanent, and occupy the northern and southern halves of the three major oceans. Their movements are dictated by the prevailing winds and the Coriolis effect. Gyres move clockwise in the Northern Hemisphere and counterclockwise in the Southern Hemisphere.

hadal zone

The deepest level of the ocean, below the abyssal zone, at depths of greater than 6,000 m/19,500 ft. The ocean trenches are in the hadal zone. There is no light in this zone and pressure is over 600 times greater than atmospheric pressure.

Hadley cell

In the atmosphere, a vertical circulation of air caused by convection. The typical Hadley cell occurs in the Tropics, where hot air over the equator in the intertropical convergence zone rises, giving the heavy rain associated with tropical rainforests. In the upper atmosphere this now dry air then spreads north and south and, cooling, descends in the latitudes of the Tropics, producing the North and South tropical desert belts. After that, the air is drawn back toward the equator, forming the Northeast and Southeast trade winds.

hail

Precipitation in the form of pellets of ice (hailstones). It is caused by the circulation of moisture in strong convection currents, usually within cumulonimbus clouds. Water droplets freeze as they are carried upward. As the circulation continues, layers of ice are deposited around the droplets until they become too heavy to be supported by the currents and they fall as a hailstorm.

halite

The mineral form of sodium chloride, NaCl. Common salt is the mineral halite. When pure it is colorless and transparent, but it is often pink, red, or yellow. It is soft and has a low density. Halite occurs naturally in evaporite deposits that have precipitated on evaporation of bodies of salt water. As rock salt, it forms beds within a sedimentary sequence; it can also migrate upward through surrounding rocks to form salt domes. It crystallizes in the cubic system.

hanging valley

A valley that joins a larger glacial trough at a higher level than the trough floor. During glaciation the ice in the smaller valley was unable to erode as deeply as the ice in the trough, and so the valley was left perched high on the side of the trough when the ice retreated. A river or stream flowing along the hanging valley often forms a waterfall as it enters the trough.

harmattan

In meteorology, a dry and dusty northeast wind that blows over West Africa.

headward erosion

The backward erosion of material at the source of a river or stream. Broken rock and soil at the source are carried away by the river, causing erosion to take place in the opposite direction to the river's flow. The resulting lowering of the land behind the source may, over time, cause the river to cut backward into a neighboring valley to "capture" another river (*see* **river capture**).

heat island

A large town or city that is warmer than the surrounding countryside. The difference in temperature is most pronounced during the winter, when the heat given off by the city's houses, offices, factories, and vehicles raises the temperature of the air by a few degrees.

hematite

The principal ore of iron, consisting mainly of iron (III) oxide, Fe_2O_3. It occurs as *specular hematite* (dark, metallic luster), *kidney ore* (reddish radiating fibers terminating in smooth, rounded surfaces), and a red earthy deposit.

hogback

A geological formation consisting of a ridge with a sharp crest and abruptly sloping sides, the outline of which resembles the back of a hog. Hogbacks are the result of differential erosion on steeply dipping rock strata composed of alternating resistant and soft beds. Exposed, almost vertical resistant beds provide the sharp crests.

Holocene

An epoch of geological time that began 10,000 years ago, the second and current epoch of the Quaternary period. During this epoch the glaciers retreated, the climate became warmer, and humans developed significantly.

horizon

The limit to which one can see across the surface of the sea or a level plain, that is, about 5 km/3 mi at 1.5 m/5 ft above sea level, and about 65 km/40 mi at 300 m/ 1,000 ft above sea level.

hornblende

A green or black rock-forming mineral, one of the amphiboles. It is a hydrous silicate composed mainly of calcium, iron, magnesium, and aluminum in addition to the silicon and oxygen that are common to all silicates. Hornblende is found in both igneous and metamorphic rocks and can be recognized by its color and prismatic shape.

hornfels

A metamorphic rock formed by rocks heated by contact with a hot igneous body. It is fine grained, brittle, and lacks foliation (a planar structure). Hornfels may contain minerals formed only under conditions of great heat, such as andalusite, Al_2SiO_5, and cordierite, $(Mg,Fe)_2Al_4Si_5O_{18}$. This rock, originating from sedimentary rock strata, is found in contact with large igneous intrusions where it represents the heat-altered equivalent of the surrounding clays. Its hardness makes it suitable for road building and railroad ballast.

hot spot

An isolated rising plume of molten mantle material that may rise to the surface of the earth's crust creating features such as volcanoes, chains of ocean islands, seamounts, and rifts in continents. Hot spots occur beneath the interiors of tectonic plates and so differ from areas of volcanic activity at plate margins (*see* **plate tectonics**). Examples of features made by hot spots are Iceland in the Atlantic Ocean, in the Pacific Ocean the Hawaiian Islands and Emperor Seamount chain, and the Galápagos Islands.

Hot spots are responsible for large amounts of volcanic activity within tectonic plates rather than at plate margins. Volcanism from a hot spot formed the unique features of Yellowstone National Park, Wyoming. The same hot spot that built Iceland atop the Mid-Atlantic ridge in the North Atlantic Ocean also produced the voluminous volcanic rocks of the Isle of Skye, Scotland, at a time before these regions were rifted apart by the opening of the Atlantic Ocean.

Chains of volcanic seamounts trace the movements of tectonic plates as they pass over hot spots. Immediately above a hot spot on oceanic crust a volcano will form. This volcano is then carried away by plate tectonic movement, and becomes extinct. A new volcano forms beside it, again above the hot spot. The result is an active volcano and a chain of increasingly old and eroded extinct volcanoes stretching away along the line traced by the plate movement. The chain of volcanoes comprising the Hawaiian Islands and Emperor Seamounts formed in this way.

Humboldt Current

The former name of the Peru Current.

humus

A component of soil consisting of decomposed or partly decomposed organic matter, dark in color and usually richer toward the surface. It has a higher carbon content than the original material and a lower nitrogen content and is an important source of minerals in soil fertility.

hurricane

Also *tropical cyclone* or *typhoon.* A severe depression (region of very low atmospheric pressure) in tropical regions, called a typhoon in the north Pacific. It is a revolving storm originating at latitudes between 5° and 20° N or S of the equator, when the surface temperature of the ocean is above 27°C/80°F. A central calm area, called the *eye,* is surrounded by inwardly spiraling winds (counterclockwise in the Northern Hemisphere) of up to 320 kph/200 mph. A hurricane is accompanied by lightning and torrential rain, and can cause extensive damage. In meteorology, a hurricane is a wind of force 12 or more on the Beaufort scale.

During 1995 the Atlantic Ocean region suffered 19 tropical storms, 11 of them hurricanes. This was the third-worst season since 1871, causing 137 deaths. The most intense hurricane recorded in the Caribbean/Atlantic sector was Hurricane Gilbert in 1988, with sustained winds of 280 kph/175 mph and gusts of over 320 kph/200 mph.

In October 1987 and January 1990 winds of near-hurricane strength were experienced in southern England. Although not technically hurricanes, they were the strongest winds there for three centuries.

The naming of hurricanes began in the 1940s with female names. Owing to public opinion that using female names was sexist, the practice was changed in 1978 to using both male and female names alternately.

hydration

A form of chemical weathering caused by the expansion of certain minerals as they absorb water. The expansion weakens the parent rock and may cause it to break up.

hydraulic action

The erosive force exerted by water (as distinct from the forces exerted by rocky particles carried by water). Hydraulic action can wear away the banks of a river, particularly at the outer curve of a meander (bend in the river), where the current flows most strongly. Hydraulic action occurs as a river tumbles over a waterfall to crash onto the rocks below. This leads to the formation of a plunge pool below the waterfall. The hydraulic action of ocean waves and turbulent currents forces air into rock cracks, and therefore brings about erosion by cavitation.

hydraulic radius

The measure of a river's channel efficiency (its ability to discharge water), used by water engineers to assess the likelihood of flooding. The hydraulic radius of a channel is defined as the ratio of its cross-sectional area to its wetted perimeter (the part of the cross section that is in contact with the water). The greater the hydraulic radius, the greater the efficiency of the channel and the less likely the river is to flood. The highest values occur when channels are deep, narrow, and semicircular in shape.

hydrograph

A graph showing how the discharge of a river varies with time. By studying hydrographs, water engineers can predict when flooding is likely and take action to prevent it. A hydrograph shows the delay between peak rainfall and the resultant peak in discharge. The shorter the time lag and the higher the peak, the more likely it is that flooding will occur. The peak flow is equal to the *groundwater flow* plus the *storm flow.* Factors likely to give short time lags and high peaks include heavy rainstorms, steep slopes, deforestation, poor soil quality, and the covering of surfaces with impermeable substances such as asphalt and concrete. Actions taken by water engineers to increase time lags and lower peaks include planting trees in the drainage basin of a river.

hydrography
The study and charting of Earth's surface waters in seas, lakes, and rivers.

hydrological cycle
An alternative name for the *water cycle,* by which water is circulated between the earth's surface and its atmosphere. It is a complex system, involving a number of physical processes (such as evaporation, precipitation, and throughflow) and stores (such as rivers, oceans, and soil).

hydrology
The study of the location and movement of inland water, both frozen and liquid, above and below ground. It is applied to major civil engineering projects such as irrigation schemes, dams, and hydroelectric power, and in planning water supply.

hydrolysis
A form of chemical weathering caused by the chemical alteration of certain minerals as they react with water. For example, the mineral feldspar in granite reacts with water to form a white clay called china clay.

hydrosphere
The water component of the earth, usually encompassing the oceans, seas, rivers, streams, swamps, lakes, groundwater, and atmospheric water vapor.

hydrothermal
Pertaining to a fluid the principal component of which is hot water, or to a mineral deposit believed to be precipitated from such a fluid.

hydrothermal vein
A crack in rock filled with minerals precipitated through the action of circulating high-temperature fluids. Igneous activity often gives rise to the circulation of heated fluids that migrate outward and move through the surrounding rock. When such solutions carry metallic ions, ore-mineral deposition occurs in the new surroundings on cooling.

hydrothermal vent
A hot fissure in the ocean floor, known as a *smoker.*

Iapetus Ocean
Also *Proto-Atlantic.* A sea that existed in early Paleozoic times between the continent that was to become Europe and that which was to become North America.

The continents moved together in the late Paleozoic, obliterating the ocean. When they moved apart once more, they formed the Atlantic.

ice age

Any period of glaciation occurring in the earth's history, but particularly that in the Pleistocene epoch, immediately preceding historic times. That particular ice age is referred to as the Ice Age. On the North American continent, glaciers reached as far south as the Great Lakes, and an ice sheet spread over northern Europe, leaving its remains as far south as Switzerland.

There were several glacial advances separated by interglacial stages during which the ice melted and temperatures were higher than today. Formerly there were thought to have been only three or four glacial advances, but recent research has shown about 20 major incidences. For example, ocean bed cores record the absence or presence in their various layers of such cold-loving small marine animals as radiolaria, which indicate a fall in ocean temperature at regular intervals. Other ice ages have occurred throughout geological time: there were four in the Precambrian era, one in the Ordovician, and one at the end of the Carboniferous and beginning of the Permian. The occurrence of an ice age is governed by a combination of factors (the Milankovitch hypothesis): (1) the earth's change of attitude in relation to the sun, that is, the way it tilts in a 41,000-year cycle and at the same time wobbles on its axis in a 22,000-year cycle, making the time of its closest approach to the sun come at different seasons; and (2) the 92,000-year cycle of eccentricity in its orbit round the sun, changing it from an elliptical to a near circular orbit, the severest period of an ice age coinciding with the approach to circularity. There is a possibility that the Pleistocene ice age is not yet over. It may reach another maximum in another 60,000 years.

The theory of ice ages was first proposed in the nineteenth century by, among others, Swiss civil engineer Ignace Venetz in 1821 and Swiss naturalist Louis Agassiz in 1837. (Before then most geologists had believed that the rocks and sediment they left behind were caused by the biblical flood.) The term *ice age* was first used by botanist Karl Schimper in 1837.

Ice Age, Little

A period of particularly severe winters that gripped northern Europe between the thirteenth and seventeenth centuries. Contemporary writings and paintings show that Alpine glaciers were much more extensive than at present. There is historical evidence that the Little Ice Age affected parts of North America as well. Rivers such as the Potomac in Washington, D.C., which do not ice over today, were so frozen that they could be used for skating.

ice sheet

A body of ice that covers a large landmass or continent; it is larger than an ice cap. During the last ice age, ice sheets spread over large parts of Europe and North

America. Today there are two ice sheets, covering much of Antarctica and Greenland. About 96 percent of all present-day ice is in the form of ice sheets. The ice sheet covering western Greenland increased in thickness by 2 m/6.5 ft between 1981 and 1993; this increase is the equivalent of a 10 percent rise in global sea levels.

iceberg

A floating mass of ice, about 80 percent of which is submerged, rising sometimes to 100 m/300 ft above sea level. Glaciers that reach the coast become extended into a broad foot; as this enters the sea, masses break off and drift toward temperate latitudes, becoming a danger to shipping.

icecap

A body of ice that is larger than a glacier but smaller than an ice sheet. Such ice masses cover mountain ranges, such as the Alps, or small islands. Glaciers often originate from icecaps.

Iceland spar

A form of calcite, $CaCO_3$, originally found in Iceland. In its pure form Iceland spar is transparent and exhibits the peculiar phenomenon of producing two images of anything seen through it. It is used in optical instruments. The crystals cleave into perfect rhombohedra.

igneous rock

A rock formed from cooling magma or lava, and solidifying from a molten state. Igneous rocks are largely composed of silica (SiO_2) and they are classified according to their crystal size, texture, method of formation, or chemical composition, for example by their proportions of light and dark minerals.

Igneous rocks that crystallize from magma below the earth's surface are called *plutonic* or *intrusive,* depending on the depth of formation. They have large crystals produced by slow cooling; examples include diabase and granite. Those extruded at the surface from lava are called *extrusive* or *volcanic.* Rapid cooling results in small crystals; basalt is an example.

ilmenite

An oxide of iron and titanium, iron titanate ($FeTiO_3$); an ore of titanium. The mineral is black, with a metallic luster. It is found as an accessory mineral in mafic igneous rocks and in sands.

impermeable rock

A rock that does not allow water to pass through it—for example, clay, shale, and slate. Unlike permeable rocks, which absorb water, impermeable rocks can support rivers. They therefore experience considerable erosion (unless, like slate, they are very hard) and commonly form lowland areas.

incised meander

In a river, a deep steep-sided meander (bend) formed by the severe downward erosion of an existing meander. Such erosion is usually brought about by the rejuvenation of a river (for example, in response to a fall in sea level).

There are several incised meanders along the middle course of the River Wye, near Chepstow, Gwent, Wales.

infiltration

The passage of water into the soil. The rate of absorption of surface water by soil (the infiltration capacity) depends on the intensity of rainfall, the permeability and compactness of the soil, and the extent to which it is already saturated with water. Once in the soil, water may pass into the bedrock to form ground water.

inselberg

Also *kopje.* German for *island mountain,* a prominent steep-sided hill of resistant solid rock, such as granite, rising out of a plain, usually in a tropical area. Its rounded appearance is caused by so-called onion-skin weathering, in which the surface is eroded in successive layers. The Sugar Loaf in Rio de Janeiro harbor in Brazil, and Ayers Rock in Northern Territory, Australia, are famous examples.

insolation

The amount of solar radiation (heat energy from the sun) that reaches the earth's surface. Insolation varies with season and latitude, being greatest at the equator and least at the Poles. At the equator the sun is consistently high in the sky: its rays strike the equatorial region directly and are therefore more intense. At the Poles the tilt of the earth means that the sun is low in the sky, and so its rays are slanted and spread out. Winds and ocean currents help to balance out the uneven spread of radiation.

insolation weathering

A type of physical or mechanical weathering that involves the alternate heating and cooling of rocks and minerals. This causes expansion and contraction, particularly of dark minerals (because they absorb more heat), setting up stresses in rocks. As rocks are poor conductors of heat, only the surface layer is affected. This may contribute to exfoliation (breaking away of the outer layer of rock).

interlocking spur

One of a series of spurs (ridges of land) jutting out from alternate sides of a river valley. During glaciation its tip may be sheared off by erosion, creating a truncated spur.

international date line (IDL)

An imaginary line that approximately follows the 180° line of longitude. The date is put forward a day when crossing the line going west, and back a day when going east. The IDL was chosen at the International Meridian Conference in 1884.

intertropical convergence zone (ITCZ)

An area of heavy rainfall found in the Tropics and formed as the trade winds converge and rise to form clouds and rain. It moves a few degrees northward during the northern summer and a few degrees southward during the southern summer, following the apparent movement of the sun. The ITCZ is responsible for most of the rain that falls in Africa. The doldrums are also associated with this zone.

intrusion

A mass of igneous rock that has formed by "injection" of molten rock, or magma, into existing cracks beneath the surface of the earth, as distinct from a volcanic rock mass that has erupted from the surface. Intrusion features include vertical cylindrical structures such as stocks, pipes, and necks; sheet structures such as dikes that cut across the strata and sills that push between them; laccoliths, which are blisters that push up the overlying rock; and batholiths, which represent chambers of solidified magma and contain vast volumes of rock.

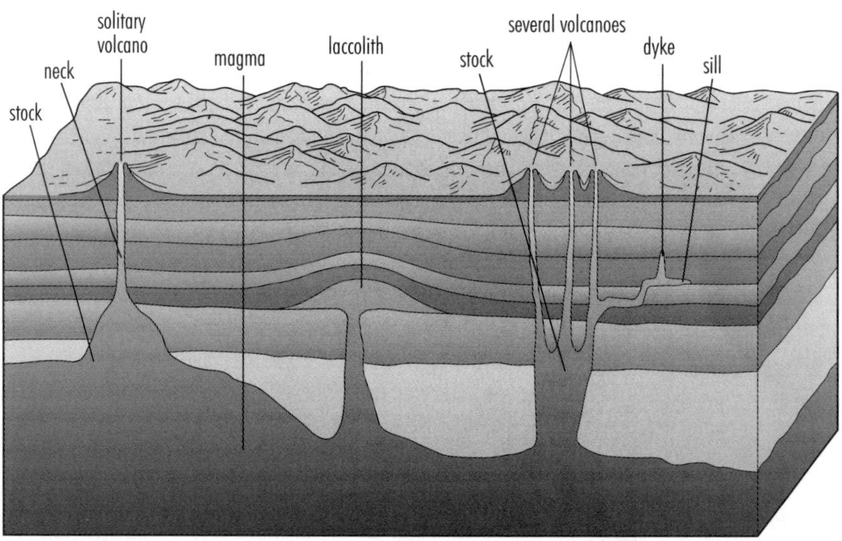

intrusive rock

An igneous rock formed beneath the earth's surface. Magma, or molten rock, cools slowly at these depths to form coarse-grained rocks, such as granite, with large crystals. (Extrusive rocks, which are formed on the surface, are usually fine grained.) A mass of intrusive rock is called an *intrusion.*

ionosphere

The ionized layer of Earth's outer atmosphere, 60–1,000 km/38–620 mi above the surface, that contains sufficient free electrons to modify the way in which radio waves are propagated—for instance by reflecting them back to Earth. The ionosphere is thought to be produced by absorption of the sun's ultraviolet radiation.

iridium anomaly

Unusually high concentrations of the element iridium found worldwide in sediments that were deposited at the Cretaceous-Tertiary boundary (K-T boundary) 65 million years ago. Because iridium is more abundant in extraterrestrial material, its presence is thought to be evidence for a large meteor impact, which may have caused the extinction of the dinosaurs. Alternatively, it may have been deposited by extensive volcanism on the continent of India during K-T time.

iron ore

Any mineral from which iron is extracted. The chief iron ores are *magnetite,* a black oxide; *hematite,* or kidney ore, a reddish oxide; *limonite,* brown, impure oxyhydroxides of iron; and *siderite,* a brownish carbonate. Iron ores are found in a number of different forms, including distinct layers in igneous intrusions, as components of contact metamorphic rocks, and as sedimentary beds. Much of the world's iron is extracted in Russia, Kazakhstan, and the Ukraine. Other important producers are the United States, Australia, France, Brazil, and Canada. Over 40 countries produce significant quantities of ore.

island

An area of land surrounded entirely by water. Australia is classed as a continent rather than an island, because of its size. Islands can be formed in many ways. *Continental islands* were once part of the mainland, but became isolated (by tectonic movement, erosion, or a rise in sea level, for example). *Volcanic islands,* such as Japan, were formed by the explosion of underwater volcanoes. *Coral islands* consist mainly of coral, built up over many years. An atoll is a circular coral reef surrounding a lagoon; atolls were formed when a coral reef grew up around a volcanic island that subsequently sank or was submerged by a rise in sea level. Barrier islands are found by the shore in shallow water, and are formed by the deposition of sediment eroded from the shoreline.

island arc

A curved chain of islands produced by volcanic activity at a destructive margin (where one tectonic plate slides beneath another). Island arcs are common in the Pacific where they ring the ocean on both sides; the Aleutian Islands off Alaska are an example. Island arcs are often later incorporated into continental margins during mountain-building episodes.

isobar

A line drawn on maps and weather charts linking all places with the same atmospheric pressure (usually measured in millibars). When used in weather forecasting, the distance between the isobars is an indication of the barometric gradient (the rate of change in pressure). Where the isobars are close together, cyclonic weather is indicated, bringing strong winds and a depression. Where isobars are far apart, the weather is anticyclonic, bringing calmer, settled conditions.

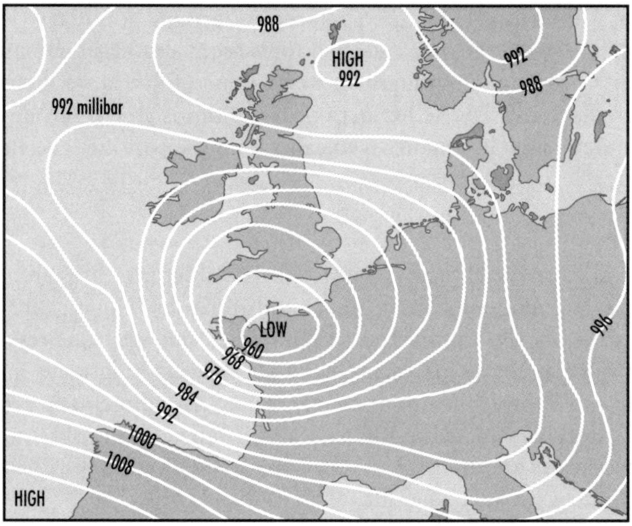

isochrone

On a map, a line that joins places that are equal in terms of the time it takes to reach them.

isohyet

On a map, a line joining points of equal rainfall.

isostasy

The theoretical balance in buoyancy of all parts of the earth's crust, as though they were floating on a denser layer beneath. There are two theories of the mechanism of isostasy, the *Airy hypothesis* and the *Pratt hypothesis,* both of which have validity. In the Airy hypothesis crustal blocks have the same density but different depths: like ice cubes floating in water, higher mountains have deeper roots. In the Pratt hypothesis, crustal blocks have different densities allowing the depth of crustal material to be the same. There appears to be more geological evidence to support the Airy hypothesis of isostasy. During an ice age, the weight of the ice sheet pushes that continent into the earth's mantle; once the ice has melted, the continent rises again. This accounts for shoreline features being found some way inland in regions that were heavily glaciated during the Pleistocene period.

isotherm

A line on a map that links all places having the same temperature at a given time.

isthmus

A narrow strip of land joining two larger landmasses. The Isthmus of Panama joins North and South America.

jacinth

Also *hyacinth.* A red or yellowish red gem, a variety of zircon, $ZrSiO_4$.

jade

A semiprecious stone consisting of either jadeite, $NaAlSi_2O_6$ (a pyroxene), or nephrite, $Ca_2(Mg,Fe)_5Si_8O_{22}(OH,F)_2$ (an amphibole), ranging from colorless through shades of green to black according to the iron content. Jade ranks 5.5–6.5 on the Mohs scale of hardness. The early Chinese civilization discovered jade, bringing it from eastern Turkestan, and carried the art of jade carving to its peak. The Olmecs, Aztecs, Maya, and the Maori have also used jade for ornaments, ceremony, and utensils.

Japan Current

Also *Kuroshio.* A warm ocean current flowing from Japan to North America.

jasper

A hard compact variety of chalcedony, SiO_2, usually colored red, brown, or yellow. Jasper can be used as a gem.

jet

A hard black variety of lignite, a type of coal. It is cut and polished for use in jewelry and ornaments. Articles made of jet have been found in Bronze Age tombs.

jet stream

A narrow band of very fast wind (velocities of over 150 kph/95 mph) found at altitudes of 10–16 km/6–10 mi in the upper troposphere or lower stratosphere. Jet streams usually occur about the latitudes of the westerlies (35°–60°). The jet stream may be used by high-flying aircraft to speed their journeys. Their discovery of the existence of the jet stream allowed the Japanese to send gas-filled balloons carrying bombs to the northwestern United States during World War II.

joint

A vertical crack in a rock, often formed by compression; it is usually several meters in length. A joint differs from a fault in that no displacement of the rocks on either side has taken place. The weathering of joints in rocks such as limestone and granite is responsible for the formation of features such as limestone pavements and tors. Joints in coastal rocks are often exploited by the sea to form erosion features such as caves and geos.

jungle

The popular name for rainforest.

Jurassic

A period of geological time 208 to 146 million years ago; the middle period of the Mesozoic era. Climates worldwide were equable, creating forests of conifers and ferns; dinosaurs were abundant, birds evolved, and limestones and iron ores were deposited. The name comes from the Jura Mountains in France and Switzerland, where the rocks formed during this period were first studied.

kame

A geological feature, usually in the form of a mound or ridge, formed by the deposition of rocky material carried by a stream of glacial meltwater. Kames are commonly laid down in front of or at the edge of a glacier (kame terrace), and are associated with the disintegration of glaciers at the end of an ice age.

Kames are made of well-sorted rocky material, usually sands and gravels. The rock particles tend to be rounded (by attrition) because they have been transported by water.

kaolin

A group of clay minerals, such as kaolinite, $Al_2Si_2O_5(OH)_4$, derived from the alteration of aluminum silicate minerals, such as feldspars and mica. It is used in medicine to treat digestive upsets, and in poultices. Kaolinite is economically important in the ceramic and paper industries. It is mined in the United States, France, the United Kingdom, and the Czech Republic.

karst

A landscape characterized by remarkable surface and underground forms, created as a result of the action of water on permeable limestone. The feature takes its name from the Karst region on the Adriatic coast in Slovenia and Croatia, but the name is applied to landscapes throughout the world, the most dramatic of which is found near the city of Guilin in the Guangxi province of China.

Limestone is soluble in the weak acid of rainwater. Erosion takes place most swiftly along cracks and joints in the limestone and these open up into gullies called *grikes.* The rounded blocks left upstanding between them are called *clints.*

katabatic wind

A cool wind that blows down a valley on calm clear nights. (By contrast, an anabatic wind is warm and moves up a valley in the early morning.) When the sky is clear, heat escapes rapidly from ground surfaces, and the air above the ground becomes chilled. The cold dense air moves downhill, forming a wind that tends to blow most strongly just before dawn. Cold air blown by a katabatic wind may collect in a depression or valley bottom to create a frost hollow. Katabatic winds are most likely to occur in the late spring and autumn because of the greater daily temperature differences.

kettle hole

A pit or depression formed when a block of ice from a receding glacier becomes isolated and buried in glacial debris (till). As the block melts, the till collapses to form a hollow, which may become filled with water to form a kettle lake or pond. Kettle holes range from 5 m/15 ft to 13 km/8 mi in diameter, and may exceed 33 m/100 ft in depth.

khamsin

A hot southeasterly wind that blows from the Sahara Desert over Egypt and parts of the Middle East from late March to May or June. It is called *sharav* in Israel.

kimberlite

An igneous rock that is ultramafic (containing very little silica); a type of alkaline peridotite with a porphyritic texture (larger crystals in a fine-grained matrix), containing mica in addition to olivine and other minerals. Kimberlite represents the world's principal source of diamonds. Kimberlite is found in carrot-shaped pipelike intrusions called *diatremes,* where mobile material from very deep in the earth's crust has forced itself upward, expanding in its ascent. The material, brought upward from near the boundary between crust and mantle, often altered and fragmented, includes diamonds. Diatremes are found principally near Kimberley, South Africa, from which the name of the rock is derived, and in the Yakut area of Siberia, Russia.

komatiite

The oldest volcanic rock with three times as much magnesium as other volcanic rocks. Unlike basaltic lavas, which comprise oceanic crust, komatiites have the chemical composition of peridotite, the primary constituent rock of the upper mantle. Komatiites were extruded as a liquid at high temperatures, perhaps more than 1,600°C/2,912°F. They have low titanium and high magnesium, nickel, and chromium content. Discovered in 1969 by Morris and Richard Viljoen, komatiites are named for the Komati River, Transvaal, South Africa, where they were first identified. The term is also used more loosely to refer to suites of other unusual basaltic rocks with which they are often associated.

K-T boundary

Geologists' shorthand for the boundary between the rocks of the Cretaceous and the Tertiary periods 65 million years ago. It coincides with the end of the extinction of the dinosaurs and in many places is marked by a layer of clay or rock enriched in the element iridium. Extinction of the dinosaurs at the K-T boundary and deposition of the iridium layer are thought to be the result of either the impact of a meteorite (or comet) that crashed into the Yucatán Peninsula (forming the Chicxulub crater) or the result of intense volcanism on the continent of India.

Kuroshio

Also *Japan Current.* A warm ocean current flowing from Japan to North America.

kyanite

Aluminum silicate, Al_2SiO_5, a pale-blue mineral occurring as blade-shaped crystals. It is an indicator of high-pressure conditions in metamorphic rocks formed from clay sediments. Andalusite, kyanite, and sillimanite are all polymorphs (*see* **polymorphism**).

laccolith

An intruded mass of igneous rock that forces apart two strata and forms a round lens-shaped mass many times wider than thick. The overlying layers are often pushed upward to form a dome. A classic development of laccoliths is illustrated in the Henry, La Sal, and Abajo Mountains of southeast Utah found on the Colorado plateau.

lagoon

A coastal body of shallow salt water, usually with limited access to the sea. The term is normally used to describe the shallow sea area cut off by a coral reef or barrier islands.

lahar

Mudflow formed of a fluid mixture of water and volcanic ash. During a volcanic eruption, melting ice may combine with ash to form a powerful flow capable of causing great destruction. The lahars created by the eruption of Nevado del Ruiz in Colombia, South America, in 1985 buried 22,000 people in 8 m/26 ft of mud.

lake

A body of still water lying in depressed ground without direct communication with the sea. Lakes are common in formerly glaciated regions, along the courses of slow rivers, and in low land near the sea. The main classifications are by origin: *glacial lakes,* formed by glacial scouring; *barrier lakes,* formed by landslides and glacial moraines; *crater lakes,* found in volcanoes; and *tectonic lakes,* occurring in natural fissures.

Crater lakes form in the calderas of extinct volcanoes, for example Crater Lake, Oregon. Subsidence of the roofs of limestone caves in karst landscape exposes the subterranean stream network and provides a cavity in which a lake can develop. Tectonic lakes form during tectonic movement, as when a rift valley is formed. Lake Tanganyika was created in conjunction with the East African Great Rift Valley. Glaciers produce several distinct types of lake, such as the lochs of Scotland and the Great Lakes of North America.

Lakes are mainly freshwater, but salt and bitter lakes are found in areas of low annual rainfall and little surface runoff, so that the rate of evaporation exceeds the rate of inflow, allowing mineral salts to accumulate. The Dead Sea has a salinity of about 250 parts per 1,000 and the Great Salt Lake, Utah, about 220 parts per 1,000. Salinity can also be caused by volcanic gases or fluids, for example Lake Natron, Tanzania.

In the twentieth century large artificial lakes have been created in connection with hydroelectric dams and other works. Some lakes have become polluted as a result of human activity. Sometimes eutrophication (a state of overnourishment) occurs when agricultural fertilizers leaching into lakes cause an explosion of aquatic life, which then depletes the lake's oxygen supply until it is no longer able to support life.

land breeze

A gentle breeze blowing from the land toward the sea and affecting coastal areas. It forms at night in the summer or autumn, and tends to be cool. By contrast, a sea breeze blows from the sea toward the land.

landslide

The sudden downward movement of a mass of soil or rocks from a cliff or steep slope. Landslides happen when a slope becomes unstable, usually because the base has been undercut or because materials within the mass have become wet and slippery. A *mudflow* happens when soil or loose material is soaked so that it no longer adheres to the slope; it forms a tongue of mud that reaches downhill from a semicircular hollow. A *slump* occurs when the material stays together as a large mass, or several smaller masses, and these may form a tilted steplike structure as they slide. A *landslip* is formed when beds of rock dipping toward a cliff slide along a lower bed. Earthquakes may precipitate landslides.

lapis lazuli

A rock containing the blue mineral lazurite in a matrix of white calcite with small amounts of other minerals. It occurs in silica-poor igneous rocks and metamorphic limestones found in Afghanistan, Siberia, Iran, and Chile. Lapis lazuli was a valuable pigment of the Middle Ages, also used as a gemstone and in inlaying and ornamental work. It was formerly used in the manufacture of ultramarine pigment.

lat.

Abbreviation for latitude. *See also* **latitude and longitude.**

lateral moraine

A linear ridge of rocky debris deposited near the edge of a glacier. Much of the debris is material that has fallen from the valley side onto the glacier's edge, having been weathered by freeze-thaw (the alternate freezing and thawing of ice in

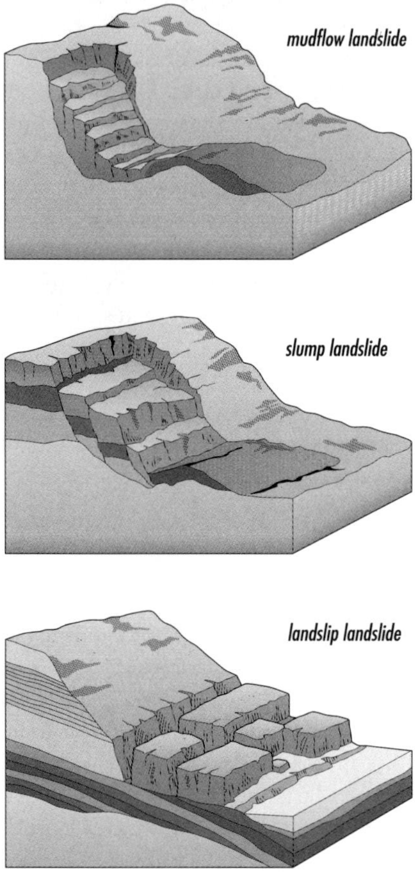

landslide

cracks); it will, therefore, tend to be angular in nature. Where two glaciers merge, two lateral moraines may join together to form a medial moraine running along the center of the merged glacier.

laterite

A red residual soil characteristic of tropical rainforests. It is formed by the weathering of basalts, granites, and shales and contains a high percentage of aluminum and iron hydroxides. It may form an impermeable and infertile layer that hinders plant growth.

latitude and longitude

Imaginary lines used to locate position on the globe. Lines of latitude are drawn parallel to the equator, with 0° at the equator and 90° at the North and South Poles. Lines of longitude are drawn at right angles to these, with 0° (the prime meridian) passing through Greenwich, England.

The 0-degree line of latitude is defined by Earth's equator, a characteristic definable by astronomical observation. It was determined as early as A.D. 150 by Egyptian astronomer Ptolemy in his world atlas. The prime meridian, or 0-degree line of longitude, is a matter of convention rather than physics. Prior to the latter half of the eighteenth century, sailors navigated by referring to their position east or west of any arbitrary meridian. When Nevil Maskelyne (1732–1811), English astronomer and fifth Astronomer Royal, published the *Nautical Almanac* he referred all of his lunar-stellar distance tables to the Greenwich meridian. These tables were relied upon for computing longitudinal position and so the Greenwich meridian became widely accepted.

Chronometers, time keeping devices with sufficient accuracy for longitude determination, invented by English instrument maker John Harrison (1693–1776) and perfected in 1759, would gradually replace the lunar distance method for navigation, but reliance on the Greenwich meridian persisted because the *Nautical Almanac* was used by sailors to verify their positions. The Greenwich meridian was officially adopted as the prime meridian by the International Meridian Conference held in Washington, D.C., in 1884.

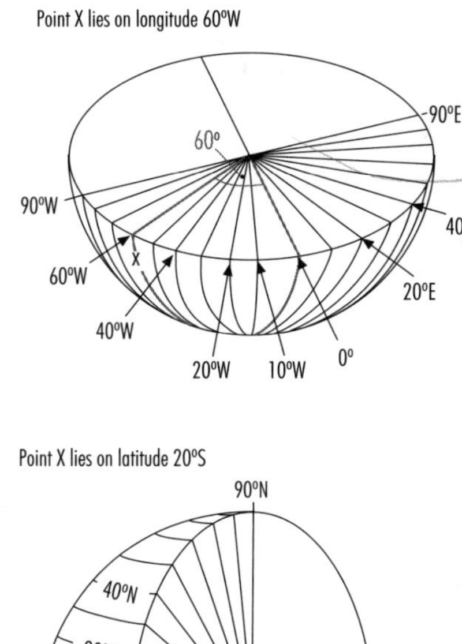

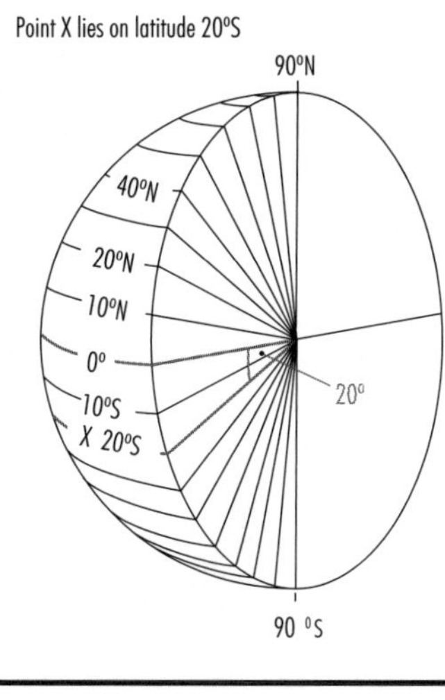

Laurasia

A northern landmass formed 200 million years ago by the splitting of the single world continent Pangaea. (The southern landmass was Gondwanaland.) It consisted of what was to become North America, Greenland, Europe, and Asia, and is believed to have broken up about 100 million years ago with the separation of North America from Europe.

lava

Molten rock (usually 800–1,100°C/1,500–2,000°F) that erupts from a volcano and cools to form extrusive igneous rock. It differs from magma in that it is molten rock on the surface; magma is molten rock below the surface. Lava that is viscous and sticky does not flow far; it forms a steep-sided conical composite volcano. Less viscous lava can flow for long distances and forms a broad flat shield volcano.

The viscosity of lava, and thus the form of volcano it forms, depends on silica content, temperature, and degree of solidification upon extrusion. It is often said that viscosity increases with silica content because silica polymerizes, but this rule can be misleading. Lavas having the composition of basalt, which is low in silica content, tend to flow easily and form broad flat volcanoes as in the Hawaiian Islands. But some very silica-rich lavas of rhyolite composition can also flow readily. Lavas that are especially viscous are often of andesite composition and intermediate in silica content. Andesite lavas can therefore give rise to explosive volcanoes like the island of Montserrat, West Indies.

The viscosity of lava was once ascribed to whether a lava was *acidic* or *basic.* These terms are misleading and no longer used. (*See* **acid rock** and **basic rock.**)

lead ore

Any of several minerals from which lead is extracted. The primary ore is galena or lead sulfite, PbS. This is unstable, and on prolonged exposure to the atmosphere it oxidizes into the minerals cerussite, $PbCO_3$, and anglesite, $PbSO_4$. Lead ores are usually associated with other metals, particularly silver, which can be mined at the same time, and zinc, which can cause problems during smelting.

Most commercial deposits of lead ore are in the form of veins, where hot fluids have leached the ore from cooling igneous masses and deposited it in cracks in the surrounding country rock, and in thermal metamorphic zones, where the heat of igneous intrusions has altered the minerals of surrounding rocks. Lead is mined in over 40 countries, but half of the world's output comes from the United States, Canada, Russia, Kazakhstan, Uzbekistan, Canada, and Australia.

leucite

A silicate mineral, $KAlSi_2O_6$, occurring frequently in some potassium-rich volcanic rocks. It is dull white to gray, and usually opaque. It is used as a source of potassium for fertilizer.

levee

A naturally formed, raised bank along the side of a river channel. When a river overflows its banks, the rate of flow is less than that in the channel, and silt is deposited on the banks. With each successive flood the levee increases in size so that eventually the river may be above the surface of the surrounding flood plain. Notable levees are found on the lower reaches of the Mississippi, and the Po in Italy.

lignite

A type of coal that is brown and fibrous, with a relatively low carbon content. As a fuel it is less efficient than bituminous coal, because more of it must be burned to produce the same amount of energy. Lignite also has a high sulfur content and is more polluting. It is burned to generate power in Scandinavia and some former eastern bloc countries because it is the only fuel resource available without importing.

limestone

A sedimentary rock composed chiefly of calcium carbonate $CaCO_3$, either derived from the shells of marine organisms or precipitated from solution, mostly in the ocean. Various types of limestone are used as building stone. Marble is metamorphosed limestone. Certain so-called marbles are not in fact marbles but fine-grained fossiliferous limestones that take an attractive polish. Caves commonly occur in limestone. Karst is a type of limestone landscape.

limestone pavement

A bare rock surface resembling a block of chocolate, found on limestone plateaus. It is formed by the weathering of limestone into individual upstanding blocks, called clints, separated from each other by joints, called grikes. The weathering process is thought to entail a combination of freeze-thaw (the alternate freezing and thawing of ice in cracks) and carbonation (the dissolving of minerals in the limestone by weakly acidic rainwater).

limnology

The study of lakes and other bodies of open fresh water, in terms of their plant and animal biology, chemistry, and physical properties.

limonite

A type of iron ore, mostly poorly crystalline iron oxyhydroxide, but usually mixed with hematite and other iron oxides. Also known as *brown iron ore,* it is often found in bog deposits.

liquefaction

The conversion of a soft deposit, such as clay, to a jellylike state by severe shaking. During an earthquake buildings and lines of communication built on materials prone to liquefaction will sink and topple. In the Alaskan earthquake of 1964 liquefaction led to the destruction of much of the city of Anchorage.

lithification

The conversion of an unconsolidated, newly deposited sediment into solid sedimentary rock by *compaction* of mineral grains that make up the sediment, new growth of the original mineral grains, and *cementation* by crystallization of new minerals from percolating aqueous solutions. The term is less commonly used to refer to solidification of molten lava or magma to form igneous rock.

lithosphere

The topmost layer of the earth's structure, forming the jigsaw of plates that take part in the movements of plate tectonics. The lithosphere comprises the crust and a portion of the upper mantle. It is regarded as being rigid and moves about on the more elastic and less rigid asthenosphere. The lithosphere is about 100 km/63 mi thick.

load

In earth sciences, material transported by a river. It includes material carried on and in the water (suspended load) and material bounced or rolled along the river bed (bedload). A river's load is greatest during a flood, when its discharge is at its highest. The term *load* can also refer to material transported by a glacier or by the sea.

loam

A type of fertile soil, a mixture of sand, silt, clay, and organic material. It is porous, which allows for good air circulation and retention of moisture.

lode

A geological deposit rich in certain minerals, generally consisting of a large vein or set of veins containing ore minerals. A system of veins that can be mined directly forms a lode, for example the mother lode of the California gold rush.

Lodes form when hot hydrothermal liquids and gases from magmas penetrate surrounding rocks, especially when these are limestones. On cooling, veins of ores formed from the magma then extend from the igneous mass into the local rock.

loess

A yellow loam, derived from glacial meltwater deposits and accumulated by wind in periglacial regions during the ice ages. Loess usually attains considerable depths, and the soil derived from it is very fertile. There are large deposits in central Europe (Hungary), China, and North America. It was first described in 1821 in the Rhine area, and takes its name from a village in Alsace.

long.

Abbreviation for longitude. *See also* **latitude and longitude.**

longitude

See **latitude and longitude.**

longshore drift

The movement of material along a beach. When a wave breaks obliquely, pebbles are carried up the beach in the direction of the wave (swash). The wave draws back at right angles to the beach (backwash), carrying some pebbles with it. In this way, material moves in a zigzag fashion along a beach. Longshore drift is responsible for the erosion of beaches and the formation of spits (ridges of sand or gravel projecting into the water). Attempts are often made to halt longshore drift by erecting barriers, or groynes, at right angles to the shore.

maelstrom

A whirlpool off the Lofoten Islands, Norway, also known as the *Moskenesstraumen,* which gave its name to whirlpools in general.

mafic rock

Plutonic rock composed chiefly of dark-colored minerals containing abundant magnesium and iron, such as olivine and pyroxene. It is derived from magnesium and ferric (iron). The term *mafic* also applies to dark-colored minerals rich in iron and magnesium as a group.

magma

Molten rock material beneath the earth's (or any of the terrestrial planets') surface from which igneous rocks are formed. Lava is magma that has extruded on to the surface.

magnetic storm

In meteorology, a sudden disturbance affecting the earth's magnetic field, causing anomalies in radio transmissions and magnetic compasses. It is probably caused by sunspot activity.

magnetite

A black, strongly magnetic opaque mineral, Fe_3O_4, of the spinel group, an important ore of iron. Widely distributed, magnetite is found in nearly all igneous and metamorphic rocks. Some deposits, called *lodestone,* are permanently magnetized. Lodestone has been used as a compass since the first millennium B.C. Today the orientations of magnetite grains in rocks are used in the study of the earth's magnetic field (*see* **paleomagnetic straigraphy**).

magnetometer

A device for measuring the intensity and orientation of the magnetic field of a particular rock or of a certain region. A magnetometer was used to confirm the phenomenon of sea-floor spreading by detecting symmetrical stripes of alternating polarity on either side of the Mid-Atlantic Ridge. In archeology, distortions of the magnetic field occur when structures, such as kilns or hearths, are present. A magnetometer allows for such features to be located without disturbing the ground, and for excavation to be concentrated in the most likely area.

malachite

A common copper ore, basic copper carbonate, $Cu_2CO_3(OH)_2$. It is a source of green pigment and is used as an antifungal agent in fish farming, as well as being polished for use in jewelry, ornaments, and art objects.

manganese ore

Any mineral from which manganese is produced. The main ores are the oxides, such as pyrolusite, MnO_2; hausmannite, Mn_3O_4; and manganite, MnO(OH). Manganese ores may accumulate in metamorphic rocks or as sedimentary deposits, frequently forming nodules on the seafloor (since the 1970s many schemes have been put forward to harvest deep-sea manganese nodules). The world's main producers are Georgia, Ukraine, South Africa, Brazil, Gabon, and India.

mangrove swamp

A muddy swamp found on tropical coasts and estuaries, characterized by dense thickets of mangrove trees. These low trees are adapted to live in creeks of salt water and send down special breathing roots from their branches to take in oxygen from the air. The roots trap silt and mud, creating a firmer, drier environment over time. Mangrove swamps are common in the Amazon delta and along the coast of West Africa, northern Australia, and Florida.

mantle

The intermediate zone of the earth between the crust and the core, accounting for 82 percent of Earth's volume. The boundary between the mantle and the crust above is the Mohorovicic discontinuity, located at an average depth of 32 km/20 mi. The lower boundary with the core is the Gutenburg discontinuity at an average depth of 2,900 km/1,813 mi.

The mantle is subdivided into *upper mantle, transition zone,* and *lower mantle,* based upon the different velocities with which seismic waves travel through these regions. The upper mantle includes a zone characterized by low velocities of seismic waves, called the *low velocity zone,* beginning at a depth of about 75 km/47 mi and extending to a depth of approximately 260 km/160 mi. This zone corresponds to the aesthenosphere upon which Earth's tectonic plates of lithosphere glide. Seismic velocities in the upper mantle are overall less than those in the transition zone and those of the transition zone are in turn less than those of the lower mantle. Faster propagation of seismic waves in the lower mantle implies that the lower mantle is more dense than the upper mantle.

The mantle is composed primarily of magnesium, silicon, and oxygen in the form of silicate minerals. In the upper mantle, the silicon in silicate minerals, such as olivine, is surrounded by four oxygen atoms. Deeper in the transition zone greater pressures promote denser packing of oxygen such that some silicon is surrounded by six oxygen atoms, resulting in magnesium silicates with garnet and pyroxene structures. Deeper still, all silicon is surrounded by six oxygen atoms so that the new mineral $MgSiO_3$, perovskite, predominates.

mantle keel

The relatively cold slab of mantle material attached to the underside of a continental craton (core of a continent composed of old, highly deformed metamorphic rock), and protruding down into the mantle like the keel of a boat. Its presence suggests that tectonic processes may have been different at the time the cratons were formed.

map

A diagrammatic representation of an area—for example, part of the earth's surface or the distribution of the stars. Modern maps of the earth are made using satellites in low orbit to take a series of overlapping stereoscopic photographs from which a three-dimensional image can be prepared. The earliest accurate large-scale maps appeared about 1580.

Conventional aerial photography, laser beams, microwaves, and infrared equipment are also used for land surveying. Many different kinds of map projection (the means by which a three-dimensional body is shown in two dimensions) are used in map making. Detailed maps requiring constant updating are kept in digital form on computer so that minor revisions can be made without redrafting.

marble

A rock formed by metamorphosis of sedimentary limestone. It takes and retains a good polish, and is used in building and sculpture. In its pure form it is white and consists almost entirely of calcite, $CaCO_3$. Mineral impurities give marble various colors and patterns. Carrara, Italy, is known for white marble.

Mariana Trench

The lowest region on the earth's surface; the deepest part of the seafloor. The trench is 2,400 km/1,500 mi long and is situated 300 km/200 mi east of the Mariana Islands, in the northwest Pacific Ocean. Its deepest part is the gorge known as the Challenger Deep, which extends 11,034 m/36,210 ft below sea level.

marl

A crumbling sedimentary rock, sometimes called *clayey limestone,* including various types of calcareous clays and fine-grained limestones. Marls are often laid down in freshwater lakes and are usually soft, earthy, and of a white, gray, or brownish color. They are used in cement making and as fertilizer.

marsh

A low-lying wetland. Freshwater marshes are common wherever groundwater, surface springs, streams, or runoff cause frequent flooding or more or less permanent shallow water. A marsh is alkaline, whereas a bog is acidic. Marshes develop on inorganic silt or clay soils. Rushes are typical marsh plants. Large marshes dominated by papyrus, cattail, and reeds, with standing water throughout the year, are commonly called swamps. Near the sea, salt marshes may form.

mass extinction

An event that produces the extinction of many species at about the same time. One notable example is the boundary between the Cretaceous and Tertiary periods (known as the K-T boundary) that saw the extinction of the dinosaurs and other big reptiles, and many of the marine invertebrates as well. Mass extinctions have taken place frequently during Earth's history. The largest mass extinction occurred approximately 245 million years ago at the boundary between the Permian period of the Paleozoic era and the Triassic period of the Mesozoic era; 90 percent of all species became extinct at that time.

mass movement

Downhill movement of surface materials under their own weight (by gravity). Some types of mass movement are very rapid—for example, landslides, whereas others, such as soil creep, are slow. Water often plays a significant role as it acts as a lubricant, enabling material to move.

meander

A loop-shaped curve in a mature river flowing sinuously across flat country. As a river flows, any curve in its course is accentuated by the current. On the outside of the curve, the velocity—and therefore the erosion—of the current is greatest. Here the river cuts into the outside bank, producing a *cutbank* or *river cliff* and the river's deepest point, or *thalweg.* On the curve's inside, the current is slow and deposits any transported material, building up a gentle slip-off slope. As each meander migrates in the direction of its cutbank, the river gradually changes its course across the flood plain.

A loop in a river's flow may become so accentuated that it becomes cut off from the normal course and forms an oxbow lake. The occurrence of meanders depends upon the gradient (slope) of the land, the nature of the river's discharge, and the type of material being carried. Meanders are common where the gradient is gentle, the discharge fairly steady (not subject to extremes), and the material carried is fine. The word *meander* comes from the River Menderes in Turkey.

mechanical weathering

An alternative name for *physical weathering.*

medial moraine

A linear ridge of rocky debris running along the center of a glacier. Medial moraines are commonly formed by the joining of two lateral moraines when two glaciers merge.

Mediterranean climate

A climate characterized by hot dry summers and warm wet winters. Mediterranean zones are situated in either hemisphere on the western side of continents, between latitudes of 30° and 60°. During the winter rain is brought by the westerlies; in summer Mediterranean zones are under the influence of the trade winds.

The regions bordering the Mediterranean Sea, California, central Chile, the Cape of Good Hope, and parts of South Australia have such climates.

meerschaum
An aggregate of minerals, usually the soft white clay mineral sepiolite, hydrous magnesium silicate. It floats on water and is used for making pipe bowls.

meltwater
Water produced by the melting of snow and ice, particularly in glaciated areas. Streams of meltwater flowing from glaciers transport rocky materials away from the ice to form outwash. Features formed by the deposition of debris carried by meltwater or by its erosive action are called fluvioglacial features; they include eskers, kames, and outwash plains.

Mercalli scale
A scale used to measure the intensity of an earthquake. It differs from the Richter scale, which measures the total amount of energy released (magnitude). It is named for the Italian seismologist Giuseppe Mercalli (1850–1914). The Mercalli scale compares the effects felt at the ground surface during an earthquake, such as windows rattling, doors swinging, or towers collapsing. Intensity is a subjective value, based on observed phenomena, and varies from place to place with the same earthquake.

meridian
Half a great circle drawn on the earth's surface passing through both Poles and thus through all places with the same longitude. Terrestrial longitudes are usually measured from the Greenwich meridian. An astronomical meridian is a great circle passing through the celestial pole and the zenith (the point immediately overhead).

mesa
Spanish for *table.* A flat-topped, steep-sided plateau, consisting of horizontal weak layers of rock topped by a resistant formation; in particular, those found in the desert areas of the United States and Mexico. A small mesa is called a *butte.*

mesosphere
A layer in the earth's atmosphere above the stratosphere and below the thermosphere. It lies between about 50 km/31 mi and 80 km/50 mi above the ground.

Mesozoic
The era of geological time 245 million to 65 million years ago, consisting of the Triassic, Jurassic, and Cretaceous periods. At the beginning of the era, the continents were joined together as Pangaea; dinosaurs and other giant reptiles dominated the sea and air; and ferns, horsetails, and cycads thrived in a warm climate worldwide. By the end of the Mesozoic era, the continents had begun to assume

their present positions, flowering plants were dominant, and many of the large reptiles and marine fauna were becoming extinct.

metamorphic rock

A rock altered in structure and composition by pressure, heat, or chemically active fluids after original formation. (If heat is sufficient to melt the original rock, technically it becomes an igneous rock upon cooling.) The term was coined in 1833 by Scottish geologist Charles Lyell (1797–1875). The mineral assemblage present in a metamorphic rock depends on the composition of the starting material (which may be sedimentary or igneous) and the temperature and pressure conditions to which it is subjected. There are two main types of metamorphism. *Thermal metamorphism,* or contact metamorphism, is brought about by the baking of solid rocks in the vicinity of an igneous intrusion (molten rock, or magma, in a crack in the earth's crust). It is responsible, for example, for the conversion of limestone to marble. *Regional metamorphism* results from the heat and intense pressures associated with the movements and collision of tectonic plates (*see* **plate tectonics**). It brings about the conversion of shale to slate, for example.

metamorphism

A geological term referring to the changes in the mineralogy and fabric of rocks in the earth caused by increasing pressure, temperature, flow of reactive fluids through the rock, or any other changes in conditions. The resulting rocks are metamorphic. All metamorphic changes take place in solid rocks. If the rocks melt and then harden, they become igneous rocks.

meteorite

A piece of rock or metal from space that reaches the surface of the earth, the moon, or other body. Most meteorites are thought to be fragments from asteroids, although some may be pieces from the heads of comets. Most are stony, although some are made of iron and a few have a mixed rock-iron composition.

Stony meteorites can be divided into two kinds: *chondrites* and *achondrites.* Chondrites contain chondrules, small spheres composed of the silicate minerals olivine and orthopyroxene as well as glass, and comprise 85 percent of all meteorites. Achondrites do not contain chondrules. Meteorites provide evidence for the nature of the solar system and may be similar to the earth's core and mantle, neither of which can be observed directly.

Thousands of meteorites hit the earth each year, but most fall in the sea or in remote areas and are never recovered. The largest known meteorite is one composed of iron, weighing 66 tons, which lies where it fell in prehistoric times at Grootfontein, Namibia.

Meteorites are slowed down by Earth's atmosphere, but if they are moving fast enough they can form a crater on impact. Meteor Crater in Arizona, about 1,200 m/4,000 ft in diameter and 200 m/650 ft deep, is the site of a meteorite impact about 50,000 years ago.

A meteorite from Mars (one of 12 so-called SNC meteorites) found in Antarctica in 1984 was revealed by NASA scientists to contain possible evidence of life in August 1996. The claim was immediately questioned by other scientists. The 4.5 billion-year-old rock entered Earth's atmosphere 12,000 years ago, probably having been broken from the surface of Mars when the planet collided with a large object in space millions of years ago. The meteorite, ALH84001, weighs 1.9 kg/4.2 lb, and was found to contain carbonate globules, on the surface of which are minute round or long and thin objects resembling microorganisms. These microfossils are a hundred times smaller than the smallest known on Earth. The globules also contain magnetite and iron sulfide particles comparable in shape and quantity to those produced by Earth microorganisms. The validity of the claim that ALH84001 contains evidence for past life on Mars is subject for debate.

meteoroid

A chunk of rock in interplanetary space. There is no official distinction between meteoroids and asteroids, except that the term *asteroid* is generally reserved for objects larger than 1.6 km/1 mi in diameter, whereas meteoroids can range anywhere from pebble size and up. Meteoroids are believed to result from the fragmentation of asteroids after collisions. Some meteoroids strike the earth's atmosphere, and their fiery trails are called *meteors.* If they fall to Earth, they are named *meteorites.*

meteorology

The scientific observation and study of the atmosphere, so that weather can be accurately forecast. Data from meteorological stations and weather satellites are collated by computer at central agencies, and forecast and weather maps based on current readings are issued at regular intervals. Modern analysis, employing some of the most powerful computers, can give useful forecasts for up to six days ahead.

At meteorological stations readings are taken of the factors determining weather conditions: atmospheric pressure, temperature, humidity, wind (using the Beaufort scale), cloud cover (measuring both the type of clouds and their coverage), and precipitation such as rain, snow, and hail (measured at 12-hour intervals). Satellites are used either to relay information transmitted from the earth-based stations, or to send pictures of cloud development, indicating wind patterns, and snow and ice cover.

Collecting data. Observations can be collected not only from land stations, but also from weather ships, aircraft, and self-recording and automatic transmitting stations, such as the radiosonde. Radar may be used to map clouds and storms. Satellites have played an important role in televising pictures of global cloud distribution.

Observation stations may be classified as follows:

> *Observatories* are where reliable standard and absolute measurements are made as far as possible with autographic instruments, which are often

duplicated for checking and research purposes. Elements such as atmospheric electricity and solar radiation are measured only at observatories and other research establishments.

Climatological reporting stations report the general daily weather conditions and make observations at standard hours during the day to provide cumulative data, such as average temperature, maximum and minimum temperatures, rainfall, sunshine, mean pressure, days of fog, frost, snowfall, and the extent and persistence of snow cover. After statistical analysis, climatic charts and tables are constructed showing the frequency of the different weather elements, such as gales and frost.

Crop weather stations make observations for use in agricultural meteorology, or micrometeorology, where the elements of the weather need to be studied in detail. It is necessary to have detailed information on temperature, humidity, and wind at heights below the average crop height in order to study and control plant diseases spread by aphids or wind-borne viruses. Details of frost hollows, wind breaks, and the degree of frost that will damage plants must all be studied.

Rainfall stations measure the amount of rain that falls. Most stations measure the daily amount, while those in remote areas measure the monthly rainfall.

Synoptic reporting stations, where observations are made simultaneously throughout the world, report in a mutually agreed form so that data can be directly compared between them. Observations are restricted to the elements required for forecasting. Reports are received at a national center and a selection is broadcast for use by other countries. The huge mass of synoptic data collected and disseminated for forecasting purposes is plotted on synoptic weather charts. Modified copies of these charts using standard symbols are published daily by most meteorological services.

Measuring and describing conditions. Meteorological observations, for whatever purpose, must be clear, precise, and strictly comparable between stations. It is easy to decide whether it is fine or cloudy, or if there is a thunderstorm; the distinctions between rain, snow, and hail are obvious; sleet is wet snow, melting snow, or a mixture of rain and snow; soft hail is halfway between snow and hail; and drizzle, which consists of very small drops, is halfway between rain and cloud, the water drops being just large enough to fall to the ground. It is useful also to describe rain as showery, intermittent, or continuous, as light, moderate, or heavy. The rate of rainfall and the total rainfall during a given period can be measured with a rain gauge. Wind strength and direction can be measured accurately by anemometers and wind vanes. Clouds are observed carefully because they are closely related to other weather conditions. Fog and other hindrances to visibility indicate approaching weather conditions as well as being of great practical importance.

Air pressure. Although the condition of the air overhead can be partly deduced from cloud observations, measurements of the physical state of the atmosphere are made at as great a height as possible. The weight of the air above any point presses downward and the force this produces in all directions is called the *air pressure;* it is measured by barometers and barographs, the unit of measurement being the millibar (mb). With increase in height there is less air above and therefore pressure decreases with height by about a factor of 10 for every increase of height of 16 km/10 mi. If the temperature of the air is known, the decrease in pressure can be calculated: near sea level it amounts to about 1 mb in every 10 m/33 ft. In order to compare pressures between many stations at a constant level, pressures are reduced to sea level, that is, barometer readings are adjusted to show what the pressure would be at sea level.

Temperature. Measuring the temperature of the air can be difficult, because a thermometer measures its own temperature, not necessarily that of its surroundings. In addition, temperature varies irregularly with height, particularly in the first few meters. Thus the temperature at 1 m/3.3 ft above the ground may easily be 5°C/41°F lower than the temperature closer to the ground, whereas on a following clear night the reverse usually occurs. On the other hand the temperature of air that rises 100 m/330 ft cools by only about 1°C/1.8°F by reason of its change of height and consequent expansion because of decrease in pressure. Temperatures are therefore read at a standard height (usually 1.2 m/3.9 ft) above the ground, and not reduced to sea level. A thermometer is kept at the same temperature as the surrounding air by sheltering it in a Stevenson screen.

Humidity. The humidity of the air is found by comparing readings taken by an ordinary thermometer with readings from a thermometer the bulb of which is covered with moist muslin. Temperature and humidity in the upper air are measured by attaching instruments to an aircraft, to a small balloon, or even to a rocket. With aircraft, measurements have been made up to more than 15 km/9.3 mi above the earth, with balloons to 30 km/18.5 mi, and with rockets to more than 150 km/93 mi. Initially, instruments carried by balloon had to be recovered before readings could be obtained; now, radiosonde balloons transmit observations to Earth. Balloons can also be used to measure winds, as they are carried by the wind, and direction-finding radio or radar enable their drift to be calculated, thus determining the speed and strength of winds at the different heights through which the balloon passes.

Weather ships. Reports from ships follow the same patterns as those from land stations, but also include the state and temperature of the sea. Weather ships are equipped to cruise at fixed points making observations, including radiosonde ascents at standard intervals. These ships are also used as navigational beacons for aircraft, and may provide rescue services if required.

mica

A group of silicate minerals that split easily into thin flakes along lines of weakness in their crystal structure (perfect basal cleavage). They are glossy, have a pearly luster, and are found in many igneous and metamorphic rocks. Their good thermal and electrical insulation qualities make them valuable in industry.

Mica's chemical composition is complicated, but micas are silicates with silicon-oxygen tetrahedra arranged in continuous sheets, with weak bonding between the layers, resulting in perfect cleavage. A common example of mica is muscovite (white mica), $KAl_2Si_3AlO_{10}(OH,F)_2$.

microclimate

The climate of a small area, such as a woodland, lake, or even a hedgerow. Significant differences can exist between the climates of two neighboring areas. For example, a town is usually warmer than the surrounding countryside (forming a heat island), and a woodland cooler, darker, and less windy than an area of open land. Microclimates play a significant role in agriculture and horticulture, as different crops require different growing conditions.

Mid-Atlantic Ridge

An ocean ridge, formed by the movement of plates described by plate tectonics, that runs along the center of the Atlantic Ocean, parallel to its edges, for some 14,000 km/8,800 mi—almost from the Arctic to the Antarctic. The

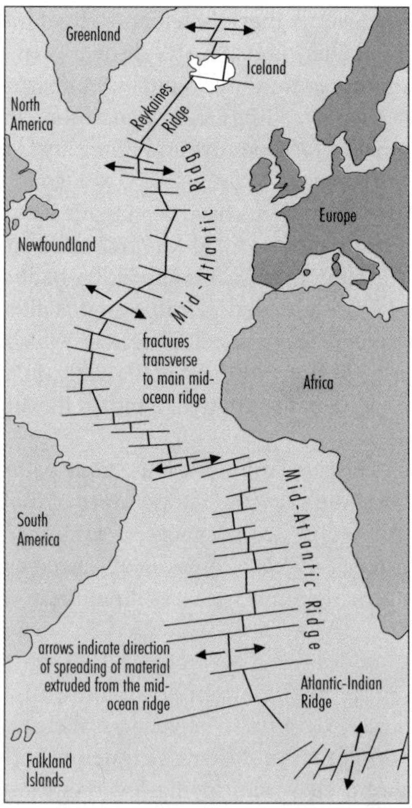

Mid-Atlantic Ridge is central because the ocean crust beneath the Atlantic Ocean has continually grown outward from the ridge at a steady rate during the past 200 million years. Iceland straddles the ridge and was formed by volcanic outpourings.

midnight sun

The constant appearance of the sun (within the Arctic and Antarctic circles) above the horizon during the summer.

Milankovitch hypothesis

The combination of factors governing climatic change and the occurrence of ice ages proposed in 1930 by the Yugoslav geophysicist M. Milankovitch (1879–1958). These include the variation in the angle of the earth's axis, and the geometry of the earth's orbit around the sun.

mineral

A naturally formed inorganic substance with a particular chemical composition and a regularly repeating internal structure. Either in their perfect crystalline form or otherwise, minerals are the constituents of rocks. In more general usage, a mineral is any substance economically valuable for mining (including coal and oil, despite their organic origins).

Mineral forming processes include: melting of preexisting rock and subsequent crystallization of a mineral to form magmatic or volcanic rocks; weathering of rocks exposed at the land surface, with subsequent transportation and grading by surface waters, ice, or wind to form sediments; and recrystallization through increasing temperature and pressure with depth to form metamorphic rocks.

Minerals are usually classified as magmatic, sedimentary, or metamorphic. The magmatic minerals include the feldspars, quartz, pyroxenes, amphiboles, micas, and olivines that crystallize from silica-rich molten rock within the crust or from extruded lavas. The most commonly occurring sedimentary minerals are either pure concentrates or mixtures of sand, clay minerals, and carbonates (chiefly calcite, aragonite, and dolomite). Minerals typical of metamorphism include andalusite, cordierite, garnet, tremolite, lawsonite, pumpellyite, glaucophane, wollastonite, chlorite, micas, hornblende, staurolite, kyanite, and diopside.

mineralogy

The study of minerals. The classification of minerals is based chiefly on their chemical composition and the kind of chemical bonding that holds these atoms together. The mineralogist also studies their crystallographic and physical characters, occurrence, and mode of formation.

The systematic study of minerals began in the eighteenth century, with the division of minerals into four classes: earths, metals, salts, and bituminous substances, distinguished by their reactions to heat and water.

Miocene

Greek for "middle recent." The fourth epoch of the Tertiary period of geological time, 23.5 to 5.2 million years ago. At this time grasslands spread over the interior of continents, and hoofed mammals rapidly evolved.

Mississippian

A term for the Lower or Early Carboniferous period of geological time, 363 to 323 million years ago. It is named for the Mississippi River valley, where rocks of this age are well exposed.

mist

A low cloud caused by the condensation of water vapor in the lower part of the atmosphere. Mist is less thick than fog, visibility being 1 to 2 km/.6 to 1.2 mi.

mistral

A cold, dry, northerly wind that occasionally blows during the winter on the Mediterranean coast of France, particularly concentrated along the Rhône valley. It has been known to reach a velocity of 145 kph/90 mph.

Mohole

A U.S. project for drilling a hole through the earth's crust, so named from the Mohorovicic discontinuity that marks the transition from crust to mantle. Initial tests were made in the Pacific in 1961, but the project was subsequently abandoned. The cores that were brought up illuminated the geological history of the earth and aided the development of geophysics.

Mohorovicic discontinuity

Also *Moho* or *M-discontinuity,* boundary that separates the earth's crust and mantle, marked by a rapid increase in the speed of earthquake waves. It follows the variations in the thickness of the crust and is found approximately 32 km/20 mi below the continents and about 10 km/6 mi below the oceans. It is named for the Yugoslav geophysicist Andrija Mohorovicic (1857–1936), who suspected its presence after analyzing seismic waves from the Kulpa Valley earthquake in 1909.

Mohs scale

A scale of hardness for minerals (in ascending order): 1 talc; 2 gypsum; 3 calcite; 4 fluorite; 5 apatite; 6 orthoclase; 7 quartz; 8 topaz; 9 corundum; 10 diamond. The scale is useful in mineral identification because any mineral will scratch any other mineral lower on the scale than itself, and similarly it will be scratched by any other mineral higher on the scale.

molybdenite

Molybdenum sulfide, MoS_2, the chief ore mineral of molybdenum. It possesses a hexagonal crystal structure similar to graphite, has a blue metallic luster, and is very soft (1–1.5 on the Mohs scale).

monazite

A mineral, $(Ce,La,Th)PO_4$, yellow to red, valued as a source of lanthanides or rare earths, including cerium and europium; generally found in placer deposit (alluvial) sands.

monsoon

A wind pattern that brings seasonally heavy rain to southern Asia; it blows toward the sea in winter and toward the land in summer. The monsoon may cause destructive flooding all over India and southeast Asia from April to September, leaving thousands of people homeless each year. The monsoon cycle is believed to have started about 12 million years ago with the uplift of the Himalayas.

moonstone

A translucent, pearly variety of potassium sodium feldspar, found in Sri Lanka and Myanmar, and distinguished by a blue, silvery, or red opalescent tint. It is valued as a gem.

moor

A stretch of land, usually at a height, which is characterized by a vegetation of heather, coarse grass, and bracken. A moor may be poorly drained and contain boggy hollows. More than 50 percent of Scotland is regarded as moorland.

moraine

A rocky debris or till carried along and deposited by a glacier. Material eroded from the side of a glaciated valley and carried along the glacier's edge is called a *lateral moraine;* that worn from the valley floor and carried along the base of the glacier is called a *ground moraine.* Rubble dropped at the snout of a melting glacier is called a *terminal moraine.* When two glaciers converge their lateral moraines unite to form a *medial moraine.* Debris that has fallen down crevasses and becomes embedded in the ice is termed an *englacial moraine;* when this is exposed at the surface due to partial melting it becomes *ablation moraine.*

mountain

A natural upward projection of the earth's surface, higher and steeper than a hill. Mountains are at least 330 m/1000 ft above the surrounding topography. The process of mountain building (orogeny) consists of thickening of the crust by folding and faulting of rock, and in some cases formation of igneous intrusions or volcanoes. Mountain building usually results from the collision of two tectonic plates (*see* **plate tectonics**) at a convergent margin. Rocks imbedded deep within mountains are subjected to high pressures and temperatures caused by the weight and heat production of the overlying rock, resulting in metamorphism. The term can also be used to describe erosional features like mesas.

mudflow

The downhill movement (mass movement) of muddy sediment containing a large proportion of water. Mudflows can be fast and destructive. In 1966 coal waste

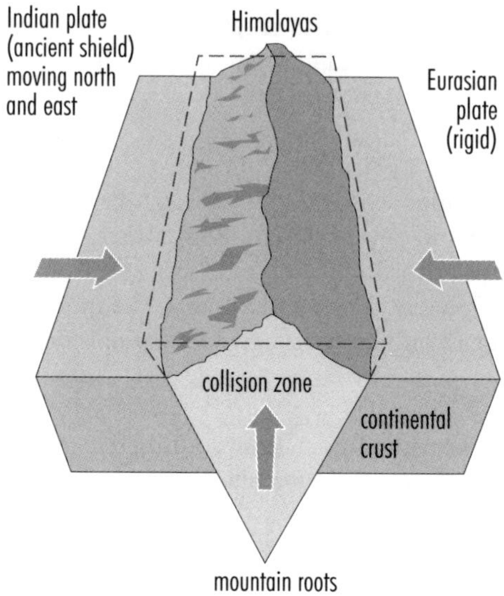

mountain

saturated with water engulfed a school in Aberfan, South Wales, killing 116 children. A *lahar* is a form of mudflow associated with volcanic activity.

mudstone

A fine-grained sedimentary rock made up of clay- to silt-sized particles (up to 0.0625 mm/0.0025 in).

muscovite

White mica, $KAl_2Si_3AlO_{10}(OH,F)_2$, a common silicate mineral. It is colorless to silvery white with shiny surfaces, and like all micas it splits into thin flakes along its one perfect cleavage. Muscovite is a metamorphic mineral occurring mainly in schists; it is also found in some granites, and appears as shiny flakes on bedding planes of some sandstones.

mylonite

A metamorphic rock composed mainly of feldspar, quartz, and micas and comprising fine-grained highly-deformed layers alternating with layers of less deformed relict grains often ovoidal in shape. The varied degree of deformation among bands produces a typically streaky appearance. They are found in areas in which rock layers have been deformed by sliding past one another parallel to their boundaries (shear strain). Once thought to be solely the result of brittle, grinding deformation (cataclasis), it is now recognized that microscopic bending followed by recrystallization has allowed portions of these rocks to deform plastically. Their unique texture can be used to reconstruct directions of movement that occurred during their formation.

native metal

Also *free metal.* Any of the metallic elements that occur in nature in the chemically uncombined or elemental form (in addition to any combined form). They include bismuth, cobalt, copper, gold, iridium, iron, lead, mercury, nickel, osmium, palladium, platinum, ruthenium, rhodium, tin, and silver. Some are commonly found in the free state, such as gold; others occur almost exclusively in the combined state, but under unusual conditions do occur as native metals, such as mercury. Examples of native nonmetals are carbon and sulfur.

New Madrid seismic fault zone

The largest system of geological faults in the eastern United States, centered on New Madrid, Missouri. Geologists estimate that there is a 50 percent chance of a magnitude 6 earthquake in the area by the year 2000. This would cause much damage because the solid continental rocks would transmit the vibrations over a wide area, and buildings in the region have not been designed with earthquakes in mind.

The zone covers a rift in the continental plate running 200 km/125 mi along the Mississippi River from northern Arkansas to southern Illinois. There are several hundred earthquakes along the fault every year, most of them too small to be felt. A series of very severe earthquakes in 1811–1812 created a lake 22 km/14 mi long and was felt as far away as Washington, D.C., and Canada.

nickel ore

Any mineral ore from which nickel is obtained. The main minerals are arsenides such as chloanthite ($NiAs_2$), and the sulfides millerite (NiS) and pentlandite $(Ni,Fe)_9S_8$, the commonest ore. The chief nickel-producing countries are Canada, Russia, Kazakhstan, Cuba, and Australia.

nivation

A complex of physical processes, operating beneath or adjacent to snow, believed to be responsible for the excavation of the hollows in which snow collects. It is also thought to play a role in the early formation of corries. The processes involved include freeze-thaw (weathering by the alternate freezing and melting of ice), mass movement (the downhill movement of substances under gravity), and erosion by meltwater.

noctilucent clouds

Clouds of ice forming in the upper atmosphere at an altitude of around 83 km/52 mi. They are visible on summer nights, particularly when sunspot activity is low.

node

In oceanography, the point on a stationary wave (an oscillating wave with no progressive motion) at which vertical motion is least and horizontal motion is greatest. In earth science, it is the point on a fault where the apparent motion has changed direction.

nodule

A lump of mineral or other matter found within rocks or formed on the seabed surface; mining technology is being developed to exploit them.

North Atlantic Drift

A warm ocean current in the North Atlantic Ocean; an extension of the Gulf Stream. It flows east across the Atlantic and has a mellowing effect on the climate of northwest Europe, particularly the British Isles and Scandinavia.

nuée ardente

A rapidly flowing, glowing white-hot cloud of ash and gas emitted by a volcano during a violent eruption. The ash and other pyroclastics in the lower part of the cloud behave like an ash flow. In 1902 a nuée ardente produced by the eruption of Mount Pelee in Martinique swept down the volcano in a matter of seconds and killed 28,000 people in the nearby town of St. Pierre.

nugget

A piece of gold found as a lump of the native metal. Nuggets occur in alluvial deposits where river-borne particles of the metal have adhered to one another.

nunatak

A mountain peak protruding through an ice sheet. Such peaks are common in Antarctica.

oasis

An area of land made fertile by the presence of water near the surface in an otherwise arid region. The occurrence of oases affects the distribution of plants, animals, and people in the desert regions of the world.

obsidian

A black or dark-colored glassy volcanic rock, chemically similar to granite, but formed by cooling rapidly on the earth's surface at low pressure. The glassy texture is the result of rapid cooling, which inhibits the growth of crystals. Obsidian was valued by the early civilizations of Mexico for making sharp-edged tools and ceremonial sculptures.

occluded front

A weather front formed when a cold front catches up with a warm front. It brings clouds and rain as air is forced to rise upward along the front, cooling and condensing as it does so.

ocean

A great mass of salt water. Strictly speaking three oceans exist on Earth—the Atlantic, Indian, and Pacific—to which the Arctic is often added. They cover approximately 70 percent or 363,000,000 sq km/140,000,000 sq mi of the total surface area of the earth. The proportion of land to water is 2:3 in the Northern Hemisphere and 1:4.7 in the Southern Hemisphere. Water levels recorded in the world's oceans have shown an increase of 10–15 cm/4–6 in over the past 100 years.

Depth. At average, Earth's oceans are 3,660 m/12,000 ft deep, but shallow ledges (continental shelves) 180 m/600 ft run out from the continents, beyond which the continental slope reaches down to the abyssal zone, the largest area, ranging from 2,000–6,000 m/6,500–19,500 ft. Only the deep-sea trenches go deeper, the deepest recorded being 11,034 m/36,201 ft (by the *Vityaz,* USSR) in the Mariana Trench of the western Pacific in 1957. The deepest sounding in the Atlantic Ocean is 9,560 m/31,365 ft in the Puerto Rico Trench. In some ocean basins the seafloor is relatively smooth, and stretches of the abyssal plain in the northwest Atlantic have been found to be flat within 2 m/6.5 ft over distances of 100 km/62 mi. Comparing the average depth of about 4 km/2.5 mi with the horizontal dimensions, which are of the order of 5,000–15,000 km/3,100–9,320 mi, gives a ratio similar to that of the width and thickness of a single sheet of paper.

Features. Deep trenches (off eastern and southeast Asia, and western South America), volcanic belts (in the western Pacific and eastern Indian Ocean), and ocean ridges (in the mid-Atlantic, eastern Pacific, and Indian Ocean) form the most salient ocean features.

Temperature. On average, temperature distribution in the oceans has three distinct layers. From the surface to a depth of usually less than 500 m/1,640 ft, the water is quite uniformly warm. The temperature decreases comparatively rapidly in a layer 500–1,000 m/1,640–3,280 ft thick to about 5°C/41°F. This region is called the *main thermocline* and beneath it lie the deep ocean waters, where temperature decreases slowly with depth. Toward higher latitudes the thermocline becomes less deep and in subpolar regions the water column is uniformly cold.

The temperature beneath the main thermocline is fairly uniform throughout the oceans, but the temperature above the thermocline depends on latitude and the predominant currents. The mean annual surface temperature in the Tropics is about 30°C/86°F; toward the Poles this may drop to -1.7°C/28.9°F, the freezing point of sea water.

Except in the Tropics the amount of heat the ocean receives at a given latitude varies with the seasons. In late spring and summer the surface temperature increases and heat is mixed downward by turbulence, to form a mixed surface layer bounded underneath by a seasonal thermocline. This mixed layer is rarely thicker than 100 m/330 ft, and the seasonal thermocline has been shown to consist of many layers several meters deep at a uniform temperature, separated by thinner regions where temperature changes rapidly with depth. During the winter the surface temperature of the layer slowly decreases and the seasonal thermocline is eroded. The annual variation in surface temperature is at most about 10°C/50°F and often less.

The capacity of tropical waters to store heat has a great influence on the global circulation of the atmosphere, because this is driven by large-scale convection in the tropical atmosphere.

Composition and salinity. Salinity averages about 3 percent; minerals commercially extracted include bromine, magnesium, potassium, and salt. Those potentially recoverable include aluminum, calcium, copper, gold, manganese, and silver.

Salinity is usually expressed as the amount of dissolved salts contained in 1,000 parts of water, with an average value of about 35 parts per thousand (ppT). Areas of particularly heavy precipitation, such as the Tropics, and those with slight evaporation or a great inflow of fresh water, have a low salinity. In the Baltic Sea, for instance, the salinity is always below 29 ppT.

Regions of the trade winds and permanent anticyclonic conditions show high salinity, but the enclosed seas (such as the Mediterranean and the Red Sea) have the highest. The most striking contrasts of salinity are only a surface feature, and are greatly reduced in deeper waters. In the case of the Dead Sea, river water has been pouring down for thousands of years into a comparatively small lake where evaporation has been consistently high, and, as a result, the very high salinity of 200 ppT has been reached.

Pressure. The pressure at any depth is due to the weight of the overlying water (and atmosphere). For every 10 m/33 ft of depth the pressure increases by about one standard atmosphere (atm), which is the pressure at sea level due to the weight of the atmosphere. Thus, at a depth of 4,000 m/13,125 ft the pressure is 400 atm, and is 1,000 atm or more at the bottom of the deepest trenches. Even at such enormous pressures marine life has been found. One of the effects of living at these great pressures is revealed when animals are brought up quickly in a trawl only to break into pieces on account of the sudden reduction in pressure. In the laboratory small unicellular creatures have been subjected to pressures as high as 600 atm without suffering any apparent harm.

Density. The density of a sample of sea water depends first on its temperature, then on its salinity, and lastly on its pressure. Increasing salinity and pressure cause an increase in density, and higher temperatures reduce density. The range of density values found in sea water is remarkably small, varying from about 1.025 at the surface to about 1.046 at 4,000 m/13,125 ft.

As density is so closely linked to temperature, the density distribution in the oceans tends to mirror the temperature distribution. Surface density is minimal in the Tropics with a value of 1.022 and increases toward higher latitudes with values of about 1.026 at 60°N or S. Vertical sections show the same three-layer structure as temperature, with a less dense, mixed upper layer lying on the thermocline that surmounts the bottom layer. In subpolar regions the density may be quite uniform throughout the water column. Small though these density differences may be, they have a profound effect on vertical motions in the ocean and large-scale circulation.

Pollution. Oceans have always been used as a dumping area for human waste, but as the quantity of waste increases, and land areas for dumping it diminish, the problem is exacerbated. Today ocean pollutants include airborne emissions from land (33 percent by weight of total marine pollution); oil from both shipping and land-based sources; toxins from industrial, agricultural, and domestic uses; sewage; sediments from mining, forestry, and farming; plastic litter; and radioactive isotopes. Thermal pollution by cooling water from power plants or other industry is also a problem, killing coral and other temperature-sensitive sedentary species.

Light. The color of the sea is a reflection in its surface of the color of the sky above. The penetration of light into the ocean is of crucial importance to marine plants, which occur as plankton. Thus they are able to survive only in the top 100 m/330 ft, or less if the water is polluted. The amount of light to be found at any depth of the ocean depends on the altitude of the sun, on the weather and surface conditions, and on the turbidity of the water (*see* **turbidity current**). Not all the colors of the spectrum are absorbed equally: red rays are quickly absorbed, but blue and violet light penetrate much further. In clear waters, such as the Sargasso Sea, violet light may be present at 150 m/500 ft deep, though at a very low strength.

An excellent example of the absorption of light by sea water is the Blue Grotto on the island of Capri, Italy, within which everything is enveloped in pure blue light. This phenomenon occurs because light entering the cave must first pass through the surface water that almost fills the entrance.

Sound. The speed at which sound travels underwater depends on density, which in turn depends on temperature, salinity, and pressure. In water at 0°C/32°F with salinity of 35 ppT and pressure of 1 atm the speed of sound is 1,445 m/4,740 ft s^{-1}. The speed of sound increases by about 4 m/13 ft s^{-1} for an increase in temperature of 1°C/33.8°F, by 1.5 m/4.9 ft s^{-1} for an increase of 1 ppT in salinity and by 1.7 m/ 5.6 ft s^{-1} for an increase in pressure of 10 atm. This dependence on density means that the distribution of sound speed displays the three vertical regions shown by temperature and density.

Below the mixed upper layer the velocity of sound decreases with depth in the region of greatest temperature change (thermocline). From about 1,000–1,500 m/3,280–4,920 ft the pressure dependence is the most important factor and the velocity of propagation increases with depth.

The variation in the speed of sound leads to the phenomenon of refraction, whereby the direction of propagation of the sound waves is altered. Thus sound waves that start in the vicinity of the minimum speed of sound at 1,000 m/3,280 ft or more tend to remain at this depth as energy spreading out in the vertical is refracted back to the original level. This gives rise to a sound channel called the *SOFAR channel.* Even quite small sound sources, for example, from whales, may be heard in the SOFAR channel at distances of many thousands of kilometers.

A similar sound channel, called the *surface sound duct,* may be formed if conditions make the speed of sound increase with depth just below the surface. Positions of maxima in the speed-of-sound profile produce shadow zones as sound energy in this region is refracted away from this depth.

Underwater sound is used to measure the ocean depth, by echo sounding, and for detection of underwater bodies such as submarines and shoals of fish. The limit on the accuracy of depth determined by an echo sounder is due to lack of knowledge of the speed of sound at each particular location, and smaller-scale variations in the mixed upper layer can cause great difficulty in the interpretation of sonar returns. (*See also* **tide.**)

ocean current

Ocean currents can be divided into two groups. Wind-driven currents are primarily horizontal and occur in the upper few hundred meters of the ocean. Thermohaline currents are caused by changes in the density of sea water due to changes in temperature and salinity. They are mainly vertical currents affecting the deep oceans. Trade winds blowing from the east in low latitudes and the westerly winds of midlatitudes along with the Coriolis effect produce a great clockwise gyre in the oceans of the Northern Hemisphere and a counterclockwise gyre in the Southern Hemisphere. The consequent return currents such as the Gulf Stream from the equator toward the Poles are relatively narrow and strong and occur on the western boundaries of the oceans.

There are also westerly countercurrents along the equator, cold northerly currents from the Arctic, and a strong circumpolar current around Antarctica, which is driven by the "roaring forties" (westerly winds). The currents in the northern Indian Ocean are more complicated and change direction with the monsoon. Some parts of the centers of the main oceanic gyres have very little current, for example the Sargasso Sea, but other areas have recently been shown to contain large eddies that are similar to, but smaller than, depressions and anticyclones in the atmosphere. The source of these eddies is uncertain but one possibility is large meanders that break off from strong western boundary currents such as the Gulf Stream. These eddies are the subject of recent research, as are the deep ocean currents, which have recently been shown to be faster than was thought.

Deep ocean currents mostly run in the reverse direction to surface currents, in that dense water sinks off Newfoundland and Tierra del Fuego and then drifts toward the equator in deep western boundary currents. In the Arctic the sinking occurs because of the cooling of Arctic water in winter, whereas in the Antarctic it is because of an increase in salinity, and hence density, due to surface seawater freezing. The return vertical flow of this thermohaline circulation occurs as a very slow rise in dense deep water toward the surface over most of the ocean. This rise is a result of winds blowing over the water that increase the depth of the wind-mixed layer and the thermocline bringing up denser water from below.

Ocean circulation and currents play a dominant role in the climate of oceanic land margins. The ocean is warmer than the land in winter and cooler in summer, so that the climate of coastal regions is equable in that annual temperature variations are smaller than in the center of large land masses, where the climate is extreme. Warm and cold ocean currents can produce climatic contrasts in places at similar latitudes, for example western Europe and eastern Canada.

Ocean Drilling Program (ODP)

Formerly the Deep-Sea Drilling Project (1966–1983), a research project initiated in the United States to sample the rocks of the ocean crust. Initially under the direction of the Scripps Institution of Oceanography, the project was planned and administered by the Joint Oceanographic Institutions for Deep Earth Sampling (JOIDES). The operation became international in 1975, when Britain, France, West Germany, Japan, and the USSR also became involved.

Boreholes were drilled in all the oceans, using the JOIDES ships *Glomar Challenger* and *Resolution.* Knowledge of the nature and history of the ocean basins was increased dramatically. The technical difficulty of drilling the seabed to a depth of 2,000 m/6,500 ft was overcome by keeping the ship in position with side-thrusting propellers and satellite navigation, and by guiding the drill using a radiolocation system. The project is intended to continue until 2005.

ocean ridge

A mountain range on the seabed indicating the presence of a constructive plate margin (where tectonic plates are moving apart and magma rises to the surface forming new crust; *see* **plate tectonics**). Ocean ridges, such as the Mid-Atlantic Ridge, consist of many segments offset along transform faults, and can rise thousands of meters above the surrounding seabed.

Ocean ridges usually have a rift valley along their crests, indicating where the flanks are being pulled apart by the growth of the plates of the lithosphere beneath. The crests are generally free of sediment; increasing depths of sediment are found with increasing distance down the flanks.

ocean trench

A deep trench in the seabed indicating the presence of a destructive margin (produced by the movements of plate tectonics). The subduction or dragging downward of one plate of the lithosphere beneath another means that the ocean floor is pulled down. Ocean trenches are found around the edge of the Pacific Ocean and the northeast Indian Ocean; minor ones occur in the Caribbean and near the Falkland Islands.

Ocean trenches represent the deepest parts of the ocean floor, the deepest being the Mariana Trench, which has a depth of 11,034 m/36,201 ft. At depths of below 6 km/3.6 mi there is no light and very high pressure; ocean trenches are inhabited by crustaceans, coelenterates (for example, sea anemones), polychaetes (a type of worm), mollusks, and echinoderms (such as sand dollars and starfish).

oceanography

The study of the oceans. Its subdivisions deal with each ocean's extent and depth, the water's evolution and composition, its physics and chemistry, the bottom topography, currents and wind effects, tidal ranges, the biology, and the various aspects of human use.

Oceanography involves the study of water movements—currents, waves, and tides—and the chemical and physical properties of the seawater. It deals with the origin and topography of the ocean floor: ocean trenches and ridges formed by plate tectonics, and continental shelves from the submerged portions of the continents. Computer simulations are widely used in oceanography to plot the possible movements of the waters, and many studies are carried out by remote sensing.

Oligocene

The third epoch of the Tertiary period of geological time, 35.5 million to 3.25 million years ago. The name, from Greek, means "a little recent," referring to the presence of the remains of some modern types of animals existing at that time.

olivenite

Basic copper arsenate, $Cu_2AsO_4(OH)$, occurring as a mineral in olive green prisms.

olivine

A greenish mineral, magnesium iron silicate, $(Mg,Fe)_2SiO_4$. It is a rock-forming mineral, present in, for example, peridotite, gabbro, and basalt. Olivine is called *peridot* when pale green and transparent, and used in jewelry.

onyx

A semiprecious variety of chalcedonic silica, SiO_2, in which the crystals are too fine to be detected under a microscope, a state known as cryptocrystalline. It has straight parallel bands of different colors: milk white, black, and red.

Sardonyx, an onyx variety, has layers of brown or red carnelian alternating with lighter layers of onyx. It can be carved into cameos.

oolite

A limestone made up of tiny spherical carbonate particles, called *ooliths*, cemented together. Ooliths have a concentric structure with a diameter up to 2 mm/0.08 in. They were formed by chemical precipitation and accumulation on ancient seafloors. The surface texture of oolites is rather like that of fish roe. The late Jurassic limestones of the British Isles are mostly oolitic in nature. The Smackover Formation of Arkansas is an example of oolitic limestone that is an important oil reservoir.

ooze

A sediment of fine texture consisting mainly of organic matter found on the ocean floor at depths greater than 2,000 m/6,600 ft. Several kinds of ooze exist, each named for its constituents. *Siliceous ooze* is composed of the silica shells of tiny marine plants (diatoms) and animals (radiolarians). *Calcareous ooze* is formed from the calcite shells of microscopic animals (foraminifera) and floating algae (coccoliths).

opal

A form of hydrous silica ($SiO_2 . nH_2O$), often occurring as stalactites and found in many types of rock. The common opal is translucent, milk white, yellow, red, blue, or green, and lustrous. Precious opal is opalescent, the characteristic play of colors being caused by close-packed silica spheres diffracting light rays within the stone.

Opal is cryptocrystalline, that is, the crystals are too fine to be detected under an optical microscope. Opals are found in Hungary; New South Wales, Australia (black opals were first discovered there in 1905); and Mexico (red fire opals).

Ordovician

The period of geological time 510 million to 439 million years ago; the second period of the Paleozoic era. Animal life was confined to the sea: reef-building algae and the first jawless fish are characteristic. The period is named for the Ordovices, an ancient Welsh people, because the system of rocks formed in the Ordovician period was first studied in Wales.

ore

A body of rock, a vein within it, or a deposit of sediment that is worth mining for the economically valuable mineral it contains. The term is usually applied to sources of metals. Occasionally metals are found uncombined (native metals), but more often they occur as compounds such as carbonates, sulfides, or oxides. The ores often contain unwanted impurities that must be removed when the metal is extracted.

Commercially valuable ores include bauxite (aluminum oxide, Al_2O_3) hematite (iron(III) oxide, Fe_2O_3), zinc blende (zinc sulfide, ZnS), and rutile (titanium dioxide, TiO_2).

Hydrothermal ore deposits are formed from fluids such as saline water passing through fissures in the host rock at an elevated temperature. Examples are the porphyry copper deposits of Chile and Bolivia, the submarine copper-zinc-iron sulfide deposits recently discovered on the East Pacific Rise, and the limestone lead-zinc deposits that occur in the southern United States and in the Pennines of Britain.

Other ores are concentrated by igneous processes, causing the ore metals to become segregated from a magma; for example, the chromite and platinum-metal-rich bands within the Bushveld, South Africa. Erosion and transportation in rivers of material from an existing rock source can lead to further concentration of heavy minerals in a deposit; for example, Malaysian tin deposits.

Weathering of rocks in situ can result in residual metal-rich soils, such as the nickel-bearing laterites of New Caledonia.

orogeny

Also *orogenesis.* The formation of mountains. It is brought about by the movements of the rigid plates making up the earth's crust and uppermost mantle (described by plate tectonics). Where two plates collide at a destructive margin rocks become folded and lifted to form chains of mountains (such as the Himalayas).

Processes associated with orogeny are faulting and thrusting (*see* **fault**), folding, metamorphism, and plutonism (*see* **plutonic rock**). However, many topographical features of mountains—cirques, U-shaped valleys—are the result of nonorogenic processes, such as weathering, erosion, and glaciation. Isostasy (uplift due to the buoyancy of the earth's crust) can also influence mountain physiography.

orographic rainfall

Rainfall that occurs when an airstream is forced to rise over a mountain range. As the air rises, it becomes cooled. The amount of moisture that air can hold decreases with decreasing temperature. So the water vapor in the rising airstream condenses, and rain falls on the windward side of the mountain. The air descending on the leeward side contains less moisture, resulting in a *rainshadow* where there is little or no rain.

Outback

The inland region of Australia. Its main inhabitants are Aborigines, miners (including opal miners), and cattle ranchers. Its harsh beauty has been recorded by such artists as Sidney Nolan.

outwash

The sands and gravels deposited by streams of meltwater (water produced by the melting of a glacier). Such material may be laid down ahead of the glacier's snout to form a large flat expanse called an *outwash plain.* Outwash is usually well sorted, the particles being deposited by the meltwater according to their size; the largest are deposited next to the snout while finer particles are deposited further downstream.

overland flow

Another term for surface runoff of water after rain.

oxbow lake

A curved lake found on the flood plain of a river. Oxbows are caused by the loops of meanders being cut off at times of flood and the river subsequently adopting a shorter course. In the southern United States, the term *bayou* is often used.

oxidation

A form of chemical weathering caused by the chemical reaction that takes place between certain iron-rich minerals in rock and the oxygen in water. It tends to result in the formation of a red-colored soil or deposit. The inside walls of canal tunnels and bridges often have deposits formed in this way.

paleomagnetic stratigraphy

The use of distinctive sequences of magnetic polarity reversals to date rocks. Magnetism retained in rocks at the time of their formation is matched with known dated sequences of polar reversals or with known patterns of secular variation.

Pangaea

Also *Pangea.* Greek for *all-land,* a single landmass made up of all the present continents, believed to have existed between 300 and 200 million years ago; the rest of the earth was covered by the Panthalassa Ocean. Pangaea split into two landmasses—Laurasia in the north and Gondwanaland in the south—which subsequently broke up into several continents. These then drifted slowly to their present positions (*see* **continental drift**).

The existence of a single "supercontinent" was first proposed by German meteorologist Alfred Wegener in 1912.

Panthalassa

An ocean that covered the surface of the earth not occupied by the world continent Pangaea between 300 million and 200 million years ago.

peat

A fibrous organic substance found in bogs and formed by the incomplete decomposition of plants such as sphagnum moss. Northern Asia, Canada, Finland, Ireland, and other places have large peat deposits, which have been dried and used as fuel from ancient times. Peat can also be used as a soil additive. Peat bogs began to be formed when glaciers retreated, about 9,000 years ago. They grow at the rate of only a millimeter a year, and large-scale digging can result in destruction both of the bog and of specialized plants growing there. The destruction of peat bogs is responsible for diminishing fish stocks in coastal waters; the runoff from the peatlands carries high concentrations of iron, which affects the growth of the plankton on which the fish feed.

pediment

A broad, gently inclined erosion surface formed at the base of a mountain as it erodes and retreats. Pediments consist of bedrock and are often covered with a thin layer of sediments, called *alluvium,* which have been eroded off the mountain.

pegmatite

An extremely coarse-grained igneous rock of any composition found in veins; pegmatites are usually associated with large granite masses.

pelagic

Of or pertaining to the open ocean, as opposed to bottom or shore areas. *Pelagic sediment* is fine-grained fragmental material that has settled from the surface waters, usually the siliceous and calcareous skeletal remains (*see* **ooze**) of marine organisms, such as radiolarians and foraminifera.

peninsula

Land surrounded on three sides by water but still attached to a larger landmass. Florida is an example.

Pennsylvanian

The term for the Upper or Late Carboniferous period of geological time, 323 million to 290 million years ago; it is named for the U.S. state, where rocks of this age are found.

peridot

A pale green, transparent gem variety of the mineral olivine.

peridotite

A rock consisting largely of the mineral olivine; pyroxene and other minerals may also be present. Peridotite is an ultramafic rock containing less than 45 percent silica by weight. It is believed to be one of the rock types making up the earth's upper mantle, and is sometimes brought from the depths to the surface by major movements, or as inclusions in lavas.

periglacial

Of or pertaining to an area bordering a glacial area but not actually covered by ice, or having similar climatic and environmental characteristics, such as mountainous areas. Periglacial areas today include parts of Siberia, Greenland, and North America. The rock and soil in these areas is frozen to a depth of several meters (permafrost) with only the top few centimeters thawing during the brief summer. The vegetation is characteristic of tundra.

During the last ice age all of southern England was periglacial. Weathering by freeze-thaw (the alternate freezing and thawing of ice in rock cracks) would have been severe, and solifluction would have taken place on a large scale, causing wet topsoil to slip from valley sides.

permafrost

The condition in which a deep layer of soil does not thaw out during the summer. Permafrost occurs under periglacial conditions. It is claimed that 26 percent of the world's land surface is permafrost. Permafrost gives rise to a poorly drained form of grassland typical of northern Canada, Siberia, and Alaska known as tundra.

permeable rock

Rock that allows water to pass through it. Rocks are permeable if they have cracks or joints running through them or if they are porous, containing many interconnected pores. Examples of permeable rocks include limestone (which is heavily jointed) and chalk (porous). Unlike impermeable rocks, which do not allow water to pass through, permeable rocks rarely support rivers and are therefore subject to less erosion. As a result they commonly form upland areas.

Permian

The period of geological time 290 million to 245 million years ago, the last period of the Paleozoic era. Its end was marked by a significant change in marine life, including the extinction of many corals and trilobites. Deserts were widespread, terrestrial amphibians and mammal-like reptiles flourished, and cone-bearing plants (gymnosperms) came to prominence. In the oceans, 49 percent of families and 72 percent of genera vanished in the late Permian. On land, 78 percent of reptile families and 67 percent of amphibian families disappeared.

perovskite

A yellow, brown, or grayish black orthorhombic mineral, $CaTiO_3$, which sometimes contains cerium. Other minerals that have a similar structure are said to have the *perovskite structure.* The term also refers to $MgSiO_3$ with the perovskite structure, the principle mineral that makes up the earth's lower mantle.

$CaTiO_3$ perovskite occurs primarily as a minor constituent of some igneous rocks in Earth's crust and mantle and in some meteorites. $MgSiO_3$ perovskite has the same chemical composition as pyroxene, a principal constituent of the upper mantle. But the extreme pressures in the lower mantle cause the oxygen atoms to be packed more tightly together than in pyroxene, giving rise to the perovskite form of $MgSiO_3$ in the lower mantle.

The perovskite structure of $MgSiO_3$ was discovered in diamond-anvil cell experiments in which upper mantle materials that are stable at the earth's surface, such as the mineral pyroxene, are squeezed at high pressures to simulate the earth's interior. Although this mineral does not occur naturally at the earth's surface it is thought to be the most abundant mineral in the mantle, and therefore Earth's most abundant mineral.

Peru Current

Formerly known as the *Humboldt Current,* a cold ocean current flowing north from the Antarctic along the west coast of South America to southern Ecuador, then west. It reduces the coastal temperature, making the western slopes of the Andes arid because winds are already chilled and dry when they meet the coast.

pervious rock

Another name for *permeable rock.*

petroleum

Also *crude oil.* A natural mineral oil, a thick greenish-brown flammable liquid found underground in permeable rocks. Petroleum consists of hydrocarbons mixed with oxygen, sulfur, nitrogen, and other elements in varying proportions. It is thought to be derived from ancient organic material that has been converted first by bacterial action and then by heat and pressure (but its origin may be chemical also).

From crude petroleum various products are made by distillation and other processes; for example, fuel oil, gasoline, kerosene, diesel oil, and lubricating oil. Petroleum products and chemicals are used in large quantities in the manufacture of detergents, artificial fibers, plastics, insecticides, fertilizers, pharmaceuticals, toiletries, and synthetic rubber.

Petroleum is formed from the remains of marine plant and animal life that existed many millions of years ago (hence it is known as a fossil fuel). Some of these remains were deposited along with rock-forming sediments under the sea where they were decomposed anaerobically (without oxygen) by bacteria, which changed the fats of the sediments into fatty acids that were in turn changed into an asphaltic material called *kerogen.* This was then converted over millions of years into petroleum by the combined action of heat and pressure. At an early stage the organic material was squeezed out of its original sedimentary mud into adjacent sandstones. Small globules of oil collected together in the pores of the

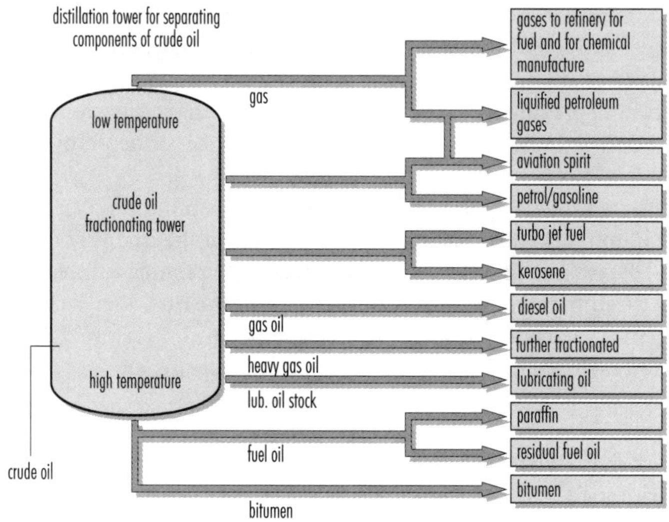

rock and eventually migrated upward through layers of porous rock by the action of the oil's own surface tension (capillary action), by the force of water movement within the rock, and by gas pressure. This migration ended either when the petroleum emerged through a fissure as a seepage of gas or oil onto the earth's surface, or when it was trapped in porous reservoir rocks, such as sandstone or limestone, in anticlines and other traps below impervious rock layers.

Production. The United States led in production until the 1960s, when the Middle East outproduced other areas, its immense reserves leading to a worldwide dependence on cheap oil for transportation and industry. In 1961 the Organization of the Petroleum Exporting Countries (OPEC) was established to avoid exploitation of member countries; after OPEC's price rises in 1973, the International Energy Agency (IEA) was established in 1974 to protect the interests of oil-consuming countries. New technologies were introduced to pump oil from offshore and from the Arctic (the Alaska pipeline) in an effort to avoid a monopoly by OPEC. Global consumption of petroleum in 1993 was 23 billion barrels.

As shallow-water oil reserves dwindle, multinational companies have been developing deep-water oil fields at the edge of the continental shelf in the Gulf of Mexico. The Shell Oil company has developed Mars, a 500-million-barrel oil field, in 900 m/2,940 ft of water, and the oil companies now have the technology to drill wells at up to 3,075 m/10,000 ft under the sea. It is estimated that the deep waters of Mexico could yield 8–15 million barrels in total and could overtake the North Sea in importance as an oil source. In Asia, the oil pipeline from Azerbaijan through Russia to the West, which is the only major pipeline from the Caspian Sea, was closed during Russia's conflict with Chechnya but reopened in 1997.

Pollution. The burning of petroleum fuel is one cause of air pollution. The transportation of oil can lead to catastrophes—for example, the *Torrey Canyon* tanker lost off southwest England in 1967, which led to an agreement by the international oil companies in 1968 to pay compensation for massive shore pollution. The 1989 oil spill in Alaska from the *Exxon Valdez* damaged the area's fragile environment, despite cleanup efforts. Drilling for oil involves the risks of accidental spillage and drilling-rig accidents. The problems associated with oil have led to the various alternative energy technologies.

A new kind of bacterium was developed during the 1970s in the United States that is capable of "eating" oil as a means of countering oil spills.

There is concern about potential damage to coral reefs as a result of future exploitation of oil reserves from the deep ocean in the North Sea off the northwest coast of Scotland, known as the *Atlantic Frontier.* Marine scientists are concerned that the North Sea oil industry will start to exploit the area before a study of the environmental impact of oil development has been completed.

petrology

The branches of geology that deal with the study of rocks, their mineral compositions, and their origins, as in igneous petrology, metamorphic petrology, and sedimentary petrology.

Phanerozoic

Greek *phanero,* meaning *visible.* An eon in Earth history, consisting of the most recent 570 million years. It comprises the Paleozoic, Mesozoic, and Cenozoic eras. The vast majority of fossils come from this eon, owing to the evolution of hard shells and internal skeletons. The name means "interval of well-displayed life."

phyllite

A metamorphic rock produced under increasing temperature and pressure, in which minute mica crystals are aligned so that the rock splits along their plane of orientation, the resulting break being shiny and smooth. An intermediate between slate and schist, its silky sheen is an identifying characteristic.

physical weathering

Also *mechanical weathering.* The form of weathering responsible for the mechanical breakdown of rocks but involving no chemical change. Processes involved include freeze-thaw (the alternate freezing and melting of ice in rock cracks or pores) and exfoliation (the alternate expansion and contraction of rocks in response to extreme changes in temperature).

pingo

A landscape feature of tundra terrain consisting of a hemispherical mound about 30 m/100 ft high, covered with soil that is cracked at the top. The core consists of ice, probably formed from the water of a former lake. The lake that forms when such a feature melts after an ice age is also called a pingo. Pingos at the Mackenzie River delta in northern Canada reach heights of up to 50 m/165 ft.

pipeflow

The movement of water through pipes in the soil. It is a form of throughflow and may be very rapid. The pipes used may be of animal or plant origin—for example, worm burrows or gaps created by tree roots.

placer deposit

A detrital concentration of an economically important mineral, such as gold, but also other minerals such as cassiterite, chromite, and platinum metals. The mineral grains become concentrated during transportation by water or wind because they are more dense than other detrital minerals such as quartz, and (like quartz) they are relatively resistant to chemical breakdown. Examples are the Witwatersrand gold deposits of South Africa, which are gold- and uranium-bearing conglomerates laid down by ancient rivers, and the placer tin deposits of the Malay Peninsula.

plain

Also *grassland.* Land, usually flat, upon which grass predominates. The plains cover large areas of the earth's surface, especially between the deserts of the Tropics and the rainforests of the equator, and have rain in one season only. In such regions the climate belts move north and south during the year, bringing rainforest conditions at one time and desert conditions at another. Temperate plains include the North European Plain, the High Plains of the United States and Canada, and the Russian Plain, also known as the *steppe.*

plate

Also *tectonic plate.* One of several sections of lithosphere approximately 100 km/60 mi thick and at least 200 km/120 mi across, which together comprise the outermost layer of the earth like the pieces of the cracked surface of a hard-boiled egg.

The plates are made up of two types of crustal material: oceanic crust (sima) and continental crust (sial), both of which are underlain by a solid layer of mantle. Dense *oceanic crust* lies beneath Earth's oceans and consists largely of basalt. *Continental crust,* which underlies the continents and their continental shelves, is thicker, less dense, and consists of rocks rich in silica and aluminum.

Due to convection in the earth's mantle (*see* **plate tectonics**) these pieces of lithosphere are in motion, riding on a more plastic layer of the mantle, called the *aesthenosphere.* Mountains, volcanoes, earthquakes, and other geological features and phenomena all come about as a result of interaction between the plates.

plate tectonics

A theory formulated in the 1960s to explain the phenomena of continental drift and sea-floor spreading, and the formation of the major physical features of the earth's surface. The earth's outermost layer, the lithosphere, is regarded as a jigsaw puzzle of rigid plates that move relative to each other, probably under the influence of convection currents in the mantle beneath. At the margins of the plates, where they collide or move apart, major landforms such as mountains, volcanoes, ocean trenches, and ocean ridges are created. The rate of plate movement is at most 15 cm/6 in per year.

The concept of plate tectonics brings together under one unifying theory many previously unrelated phenomena. The size of the lithospheric plates is variable, as they are constantly changing, but six or seven large plates now cover much of the earth's surface, the remainder being occupied by a number of smaller plates. Each plate consists of brittle mantle overlain by oceanic or continental crust. Both types of crust may be present in a single plate. As a result of seismic studies it is known that the lithosphere is a rigid layer extending to depths of 50–100 km/30–60 mi. Beneath the plates lies the asthenosphere, consisting of mantle rocks with low shear strength. The lack of shear strength in the asthenosphere results from the high temperature of the rocks, approaching the melting point. This zone of mechanical weakness allows the movement of the overlying lithospheric plates. The margins of the plates are defined by major earthquake zones and belts of volcanic and tectonic activity, which have been well known for many years. Almost all earthquake, volcanic, and tectonic activity is confined to the margins of plates, and shows that the plates are in constant motion.

Constructive margins. Where two plates are moving apart from each other, molten rock from the mantle wells up in the space between the plates and hardens to form new crust and underlying mantle, usually in the form of an ocean ridge (such as the Mid-Atlantic Ridge). The newly formed crust and mantle, or lithosphere, accumulates at the ocean ridge, adding to the plates on either side of the ridge and thereby causing the seafloor to spread; the floor of the Atlantic Ocean is growing by 5 cm/2 in each year because of the welling up of new material at the Mid-Atlantic Ridge.

Destructive margins. Where two plates are moving toward each other, the denser of the two plates may be forced to dive below the other, forming a subduction zone. The descending plate melts to form bodies of magma that may then rise to the surface through cracks and faults to form a series of volcanoes, called a *volcanic arc.* If the two plates of lithosphere are capped by buoyant continental crust, subduction of the crust does not occur. Instead it crumples gradually to form mountains, such as the Himalayas, the Andes in South America, and the Rockies in North America. This process of mountain building is termed *orogenesis.*

Conservative margins. Sometimes two plates will slide past each other. An example is the San Andreas fault, California, where the movement of the plates sometimes takes the form of sudden jerks, causing the earthquakes common in the San Francisco–Los Angeles area. Most of the earthquake zones of the world are found in regions where two plates meet or are moving apart.

Causes of plate movement. The causes of plate movement are poorly understood. It has been known for some time that heat flow from the interior of the earth is high over the ocean ridges, and so thermal convection in the mantle has been proposed as a driving mechanism for the plates. The rising limbs of the

sea-floor spreading

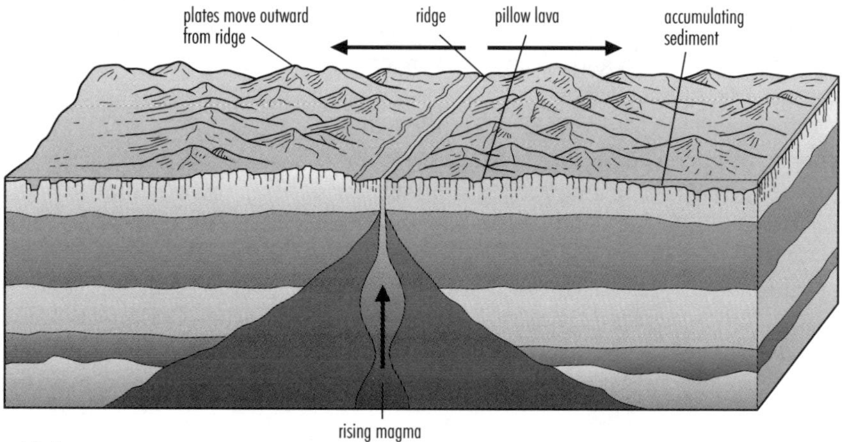

subduction zone

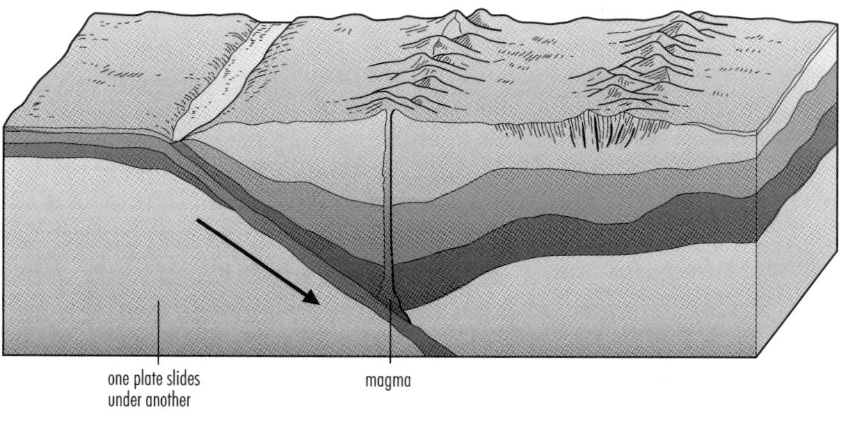

collision zone

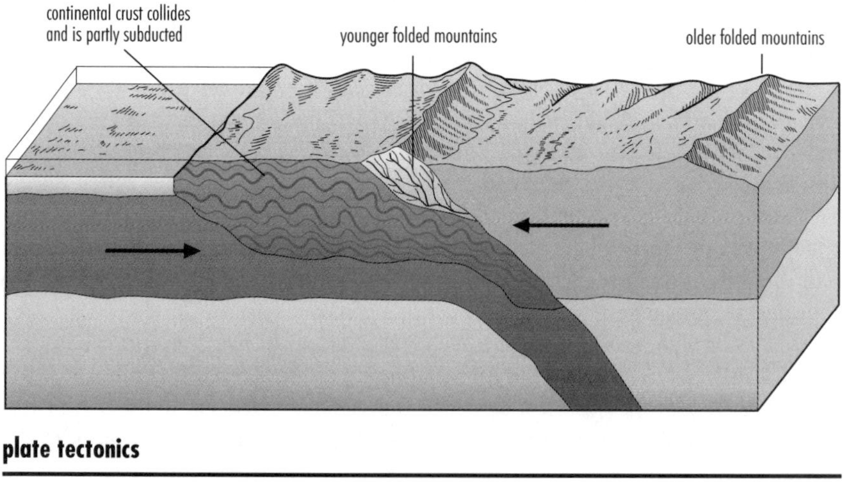

plate tectonics

convective mantle cells may be the plumes of hot, molten material that rise beneath the ocean ridges to be extruded as basaltic lava. The descending limbs of the convective cells may be linked to subduction zones. Ocean crust is continually produced by, and returned to, the mantle, but the continental crust rocks, because of their buoyancy, remain on the surface.

plateau

An elevated area of fairly flat land, or a mountainous region in which the peaks are at the same height. An *intermontane plateau* is one surrounded by mountains. A *piedmont plateau* is one that lies between the mountains and low-lying land. A *continental plateau* rises abruptly from low-lying lands or the sea. Examples are the Tibetan Plateau and the Massif Central in France.

playa

A temporary lake in a region of interior drainage. Such lakes are common features in arid desert basins, such as those in the Mojave Desert in the southwest United States, fed by intermittent streams. The streams bring dissolved salts to the lakes, and when the lakes shrink during dry spells, the salts precipitate as evaporite deposits.

Pleistocene

The first epoch of the Quaternary period of geological time, beginning 1.64 million years ago and ending 10,000 years ago. The polar ice caps were extensive and glaciers were abundant during the ice age of this period, and humans evolved into modern *Homo sapiens sapiens* about 100,000 years ago.

Pliocene

Greek for "almost recent." The fifth and last epoch of the Tertiary period of geological time, 5.2 million to 1.64 million years ago. The earliest hominid, the humanlike ape *Australopithecines,* evolved in Africa during this time.

plucking

A process of glacial erosion. Water beneath a glacier freezes fragments of loose rock to the base of the ice. When the ice moves, the rock fragment is "plucked" away from the underlying bedrock. Plucking is thought to be responsible for the formation of steep, jagged slopes such as the backwall of the corrie and the downslope side of the roche moutonnée.

plunge pool

A deep pool at the bottom of a waterfall. It is formed by the hydraulic action of the water as it crashes down onto the river bed from a height.

plutonic rock

An igneous rock derived from magma that has cooled and solidified deep in the crust of the earth; granites and gabbros are examples of plutonic rocks.

podzol

Also *podsol.* A type of light-colored soil found predominantly under coniferous forests and on moorlands in cool regions where rainfall exceeds evaporation. The constant downward movement of water leaches nutrients from the upper layers, making podzols poor agricultural soils. The leaching of minerals such as iron, lime, and alumina leads to the formation of a bleached zone, often also depleted of clay. These minerals can accumulate lower down the soil profile to form a hard impermeable layer that restricts the drainage of water through the soil.

polar reversal

A change in polarity of Earth's magnetic field. Like all magnets, Earth's magnetic field has two opposing regions, or poles, one of attraction and one of repulsion, positioned approximately near the geographical North and South Poles. During a period of normal polarity the region of attraction corresponds with the North Pole. Today, a compass needle, like other magnetic materials, aligns itself parallel to the magnetizing force and points to the North Pole. During a period of reversed polarity, the region of attraction would change to the South Pole and the needle of a compass would point south.

Studies of the magnetism retained in rocks at the time of their formation (like little compasses frozen in time) have shown that the polarity of the magnetic field has reversed repeatedly throughout geological time.

Polar reversals are a random process. Although the average time between reversals over the last 10 million years has been 250,000 years, the rate of reversal has changed continuously over geological time. The most recent reversal was 700,000 years ago; scientists have no way of predicting when the next reversal will occur. The reversal process takes about a thousand years. Movements of Earth's molten core are thought to be responsible for the magnetic field and its polar reversals. Dating rocks using distinctive sequences of magnetic reversals is called paleomagnetic stratigraphy.

polder

An area of flat reclaimed land that used to be covered by a river, lake, or the sea. Polders have been artificially drained and protected from flooding by building dikes. They are common in the Netherlands, where the total land area has been increased by nearly one-fifth since A.D. 1200. Such schemes as the Zuider Zee project have provided some of the best agricultural land in the country.

pole

Either of the geographic north and south points of the axis about which the earth rotates. The geographic poles differ from the magnetic poles, which are the points toward which a freely suspended magnetic needle will point. The changes in the position of Earth's magnetic poles measured with respect to the geographical positions is called *secular variation.*

In 1985 the magnetic North Pole was some 350 km/218 mi northwest of Resolute Bay, Northwest Territories, Canada. It moves northward about 10 km/6 mi each year, although it can vary in a day about 80 km/50 mi from its average position. It is relocated every decade in order to update navigational charts.

It is thought that periodic changes in the earth's core cause a reversal of the magnetic poles (*see* **polar reversal**). Many animals, including migrating birds and fish, are believed to orient themselves partly using the earth's magnetic field. A permanent scientific base collects data at the South Pole.

polymorphism

In mineralogy, the ability of a substance to adopt different internal structures and external forms in response to different conditions of temperature and/or pressure. For example, diamond and graphite are both forms of the element carbon, but they have very different properties and appearance. Silica (SiO_2) also has several polymorphs, including quartz, tridymite, cristobalite, and stishovite (the latter a very high-pressure form found in meteoritic impact craters).

pools and riffles

Alternating deeps (pools) and shallows (riffles) along the course of a river. There is evidence to suggest a link between pools and riffles and the occurrence of meanders (bends in a river), although it is not certain whether they are responsible for meander formation.

porphyry

Any igneous rock containing large crystals in a finer matrix.

pothole

A small hollow in the rock bed of a river. Potholes are formed by the erosive action of rocky material carried by the river (corrasion), and are commonly found along the river's upper course, where it tends to flow directly over solid bedrock.

pozzolan

Also *pozzolana* or *puzzolan*. A silica-rich material, such as volcanic tuff or chert, that can be ground up and mixed to form a cement highly resistant to corrosion by salt water. Currently used in portland cement, it was named for the town of Pozzuoli, Italy, where a nearby volcanic tuff was used as a basis for cement in Roman times.

prairie

The central North American plain, formerly grass covered, extending over most of the region between the Rocky Mountains on the west and the Great Lakes and Ohio River on the east.

Precambrian

The time from the formation of Earth (4.6 billion years ago) up to 570 million years ago. Its boundary with the succeeding Cambrian period marks the time when animals first developed hard outer parts (exoskeletons) and so left abundant fossil remains. It comprises about 85 percent of geological time and is divided into two periods: the archean, in which no life existed, and the Proterozoic, in which there was life in some form.

precipitation

In meteorology, water that falls to the earth from the atmosphere. It is part of the hydrological cycle. Forms of precipitation include rain, snow, sleet, hail, dew, and frost. The amount of precipitation in any one area depends on climate, weather, and phenomena like trade winds and ocean currents. The cyclical change in the Peru Current off the coasts of Ecuador and Peru, known as *El Niño,* causes dramatic shifts in the amount of precipitation in South and Central America and throughout the Pacific region.

Precipitation can also be influenced by people. In urban areas dust, smoke, and other particulate pollution that comprise *condensation nuclei* cause water in the air to condense more readily. Fog is one example. Precipitation also can react chemically with air-borne pollutants to produce acid rain.

prevailing wind

The direction from which the wind most commonly blows in a locality. In northwest Europe, for example, the prevailing wind is southwesterly, blowing from the Atlantic Ocean in the southwest and bringing moist and warm conditions.

primary data

Information that has been collected firsthand. It involves measurement of some sort, whether by taking readings off instruments, sketching, counting, or conducting interviews (using questionnaires).

Proterozoic

The eon of geological time, 3.5 billion to 570 million years ago, the second division of the Precambrian. It is defined as the time of simple life, because many rocks dating from this eon show traces of biological activity, and some contain the fossils of bacteria and algae.

provenance

The area from which sedimentary materials are derived.

pseudomorph

A mineral that has replaced another *in situ* and has retained the external crystal shape of the original mineral.

pumice

A light volcanic rock produced by the frothing action of expanding gases during the solidification of lava. It has the texture of a hard sponge and is used as an abrasive.

pyramidal peak

An angular mountain peak with concave faces found in glaciated areas; for example, the Matterhorn in Switzerland. It is formed when three or four corries (steep-sided hollows) are eroded, back-to-back, around the sides of a mountain, leaving an isolated peak in the middle.

pyrite

Iron sulfide, FeS_2, also called *fool's gold* because of its yellow metallic luster. Pyrite has a hardness of 6–6.5 on the Mohs scale. It is used in the production of sulfuric acid.

pyroclastic

Of or pertaining to fragments of solidified volcanic magma, ranging in size from fine ash to large boulders, that are extruded during an explosive volcanic eruption; also the rocks that are formed by consolidation of such material. Pyroclastic rocks include tuff (ash deposit) and agglomerate (volcanic breccia).

pyroclastic deposit

A deposit made up of fragments of rock, ranging in size from fine ash to large boulders, ejected during an explosive volcanic eruption.

pyroxene

Any one of a group of minerals, silicates of calcium, iron, and magnesium with a general formula X,YSi_2O_6, found in igneous and metamorphic rocks. The internal structure is based on single chains of silicon and oxygen. Diopside (X = Ca, Y = Mg) and augite (X = Ca, Y = Mg,Fe,Al) are common pyroxenes. Jadeite ($NaAlSi_2O_6$), which is considered the more valuable form of jade, is also a pyroxene.

quartz

A crystalline form of silica, SiO_2, one of the most abundant minerals of the earth's crust (12 percent by volume). Quartz occurs in many different kinds of rock, including sandstone and granite. It ranks 7 on the Mohs scale of hardness and is resistant to chemical or mechanical breakdown. Quartzes vary according to the size and purity of their crystals. Crystals of pure quartz are coarse, colorless, transparent, show no cleavage, and fracture unevenly; this form is usually called *rock crystal.* Impure colored varieties, often used as gemstones, include agate, citrine quartz, and amethyst. Quartz is also used as a general name for the cryptocrystalline and noncrystalline varieties of silica, such as chalcedony, chert, and opal.

Quartz is used in ornamental work and industry, where its reaction to electricity makes it valuable in electronic instruments. Quartz can also be made synthetically. Crystals that would take millions of years to form naturally can now be "grown" in pressure vessels to a standard that allows them to be used in optical and scientific instruments and in electronics, such as quartz wristwatches.

quartzite

A metamorphic rock consisting of pure quartz sandstone that has recrystallized under increasing heat and pressure.

In sedimentology, quartzite may also be an unmetamorphosed sandstone composed chiefly of quartz grains held together by silica that was precipitated after the original sand was deposited.

Quaternary

The period of geological time that began 1.64 million years ago and is still in process. It is divided into the Pleistocene and Holocene epochs.

rain

A form of precipitation in which separate drops of water fall to the earth's surface from clouds. The drops are formed by the accumulation of fine droplets that condense from water vapor in the air. The condensation is usually brought about by rising and subsequent cooling of air. Rain can form in three main ways: (1) frontal (or cyclonic) rainfall, (2) orographic (or relief) rainfall, and (3) convectional rainfall. *Frontal rainfall* takes place at the boundary, or front, between a mass of warm air from the tropics and a mass of cold air from the Poles. The water vapor in the warm air is chilled and condenses to form clouds and rain.

Orographic rainfall occurs when an airstream is forced to rise over a mountain range. The air becomes cooled and precipitation takes place. In the United Kingdom, the Pennine hills, which extend southward from Northumbria to Derbyshire in northern England, interrupt the path of the prevailing southwesterly winds, causing orographic rainfall. Their presence is partly responsible for the west of the British Isles being wetter than the east. *Convectional rainfall,* associated with hot climates, is brought about by rising and abrupt cooling of air that has been warmed by the extreme heat of the ground surface. The water vapor carried by the air condenses and so rain falls heavily. Convectional rainfall is usually accompanied by a thunderstorm, and it can be intensified over urban areas due to higher temperatures (*see* **heat island**).

rainfall gauge

An instrument used to measure precipitation, usually rain. It consists of an open-topped cylinder, inside which there is a close-fitting funnel that directs the rain to a collecting bottle inside a second, inner cylinder. The gauge may be partially embedded in soil to prevent spillage. The amount of water that collects in the bottle is measured every day, usually in millimeters.

When the amount of water collected is too little to be measured, *trace rainfall* is said to have taken place. Snow falling into the gauge must be melted before a measurement is taken.

rainforest

A dense forest usually found on or near the equator where the climate is hot and wet. Moist air brought by the converging trade winds rises because of the heat producing heavy rainfall. Over half the tropical rainforests are in Central and South America, primarily the lower Amazon and the coasts of Ecuador and Columbia. The rest are in southeast Asia (Malaysia, Indonesia, and New Guinea) and in West Africa and the Congo.

Tropical rainforest once covered 14 percent of the earth's land surface, but are now being destroyed at an increasing rate as their valuable timber is harvested and the land cleared for agriculture, causing problems of deforestation. Although by 1991 over 50 percent of the world's rainforests had been removed, they still comprise about 50 percent of all growing wood on the planet, and harbor at least 40 percent of the earth's species of plants and animals.

The vegetation in tropical rainforests typically includes an area of dense forest called *selva;* a *canopy* formed by high branches of tall trees providing shade for lower layers; an intermediate layer of shorter trees and tree roots; *lianas* (vines); and a ground cover of mosses and ferns. The lack of a seasonal rhythm causes adjacent plants to flower and shed leaves simultaneously. Chemical weathering and leaching take place in the iron-rich soil due to the high temperatures and humidity.

Rainforests comprise some of the most complex and diverse ecosystems on the planet, deriving their energy from the sun and photosynthesis. The trees are the main producers. Herbivores such as insects, caterpillars, and monkeys feed on the plants and trees and in turn are eaten by the carnivores, such as ocelots and puma. Fungi and bacteria, the primary decomposers, break down the dead material from the plants, herbivores, and carnivores with the help of heat and humidity. This decomposed material provides the nutrients for the plants and trees.

The rainforest ecosystem helps to regulate global weather patterns—especially by taking up CO_2 (carbon dioxide) from the atmosphere—and stabilizes the soil. Rainforests provide the bulk of the oxygen needed for plant and animal respiration. When deforestation occurs, the microclimate of the mature forest disappears; soil erosion and flooding become major problems because rainforests protect the shallow tropical soils. Once an area is cleared it is very difficult for shrubs and bushes to reestablish because soils are poor in nutrients. This causes problems for plans to convert rainforests into agricultural land. After two or three years the crops fail and the land is left bare. Clearing of the rainforests may lead to a global warming of the atmosphere, and contribute to the greenhouse effect.

Tropical rainforests are characterized by a great diversity of species, usually of tall broad-leafed evergreen trees, with many climbing vines and ferns, some of which are a main source of raw materials for medicines. A tropical forest, if properly preserved, can yield medicinal plants, oils (from cedar, juniper, cinnamon, and sandalwood), spices, gums, resins (used in inks, lacquers, and linoleum), tanning and dyeing materials, forage for animals, beverages, poisons, green manure, rubber, and animal products (feathers, hides, and honey). Other rainforests include montane, upper montane or cloud, mangrove, and subtropical.

Traditional ways of life in tropical rainforests are disappearing. The practice of shifting cultivation, in which small plots of forest are cultivated and abandoned after two or three harvests, is being replaced by slash-and-burn cultivation on such a large scale that the rainforests cannot regenerate. As a result hunting and gathering as a way of life also is becoming less viable. In the last 30 years, Central America has lost almost two-thirds of its rainforests to cattle ranching.

raised beach

A beach that has been raised above the present-day shoreline and is therefore no longer washed by the sea. It is an indication of a fall in sea level (eustatic) or of a rise in land level (isostatic).

Raman effect

The scattering of light accompanied by shifts in the wavelength of the light; the Raman effect is useful for probing the structure of minerals.

regolith

The surface layer of loose material that covers most bedrock. It consists of eroded rocky material, volcanic ash, river alluvium, vegetable matter, or a mixture of these, known as *soil.*

rejuvenation

The renewal of a river's powers of downward erosion. It may be caused by a fall in sea level or a rise in land level, or by the increase in water flow that results when one river captures another (river capture).

Several river features are formed by rejuvenation. For example, as a river cuts down into its channel it will leave its old flood plain perched up on the valley side to form a river terrace. Meanders (bends in the river) become deeper and their sides more steep, forming incised meanders, and waterfalls and rapids become more common.

remote imaging

Photographing the earth's surface with orbiting satellites. With a simple aerial, receiver, and software it is possible to download images straight onto a personal computer—helping amateur meteorologists, for example, to make weather forecasts.

resources

Materials that can be used to satisfy human needs. Because human needs are diverse and extend from basic physical requirements, such as food and shelter, to ill-defined aesthetic needs, resources encompass a vast range of items. The intellectual resources of a society—its ideas and technologies—determine which aspects of the environment meet that society's needs, and therefore become resources. For example, in the nineteenth century, uranium was used only in the manufacture of colored glass. Today, with the advent of nuclear technology, it is a military and energy resource. Resources are often categorized into *human re-*

sources, such as labor, supplies, and skills, and *natural resources,* such as climate, fossil fuels, and water. Natural resources are divided into nonrenewable resources and renewable resources.

Nonrenewable resources include minerals such as coal, copper ores, and diamonds, which exist in strictly limited quantities. Once consumed they will not be replenished within the time span of human history. In contrast, water supplies, timber, food crops, and similar resources can, if managed properly, provide a steady yield virtually forever; they are therefore replenishable or renewable resources. Inappropriate use of renewable resources can lead to their destruction, as for example the cutting down of rainforests, with secondary effects, such as the decrease in oxygen and the increase in carbon dioxide and the ensuing greenhouse effect. Some renewable resources, such as wind or solar energy, are continuous; supply is largely independent of human actions.

Demands for resources made by rich nations are causing concern that the present and future demands of industrial societies cannot be sustained for more than a century or two, and that this will be at the expense of the Third World and the global environment. Other authorities believe that new technologies will emerge, enabling resources that are now of little importance to replace those being exhausted.

rhyolite

An igneous rock, the fine-grained volcanic (extrusive) equivalent of granite.

ria

A long narrow sea inlet, usually branching and surrounded by hills. A ria is deeper and wider toward its mouth, unlike a fjord. It is formed by the flooding of a river valley due to either a rise in sea level or a lowering of a landmass.

ribbon lake

A long, narrow lake found on the floor of a glacial trough. A ribbon lake will often form in an elongated hollow carved out by a glacier, perhaps where it came across a weaker band of rock. Ribbon lakes can also form when water ponds up behind a terminal moraine or a landslide.

Richter scale

A scale based on measurement of seismic waves, used to determine the magnitude of an earthquake at its epicenter. The magnitude of an earthquake differs from its intensity, measured by the Mercalli scale, which is subjective and varies from place to place for the same earthquake. The scale is named for U.S. seismologist Charles Richter.

An earthquake's magnitude is a function of the total amount of energy released, and each point on the Richter scale represents a thirtyfold increase in energy over the previous point. The greatest earthquake ever recorded, in 1920 in Gansu, China, measured 8.6 on the Richter scale.

ridge of high pressure

An elongated area of high atmospheric pressure extending from an anticyclone. On a synoptic weather chart it is shown as a pattern of lengthened isobars. The weather under a ridge of high pressure is the same as that under an anticyclone.

rift valley

A valley formed by the subsidence of a block of the earth's crust between two or more parallel faults. Rift valleys are steep sided and form where the crust is being pulled apart, as at ocean ridges, or in the Great Rift Valley of East Africa.

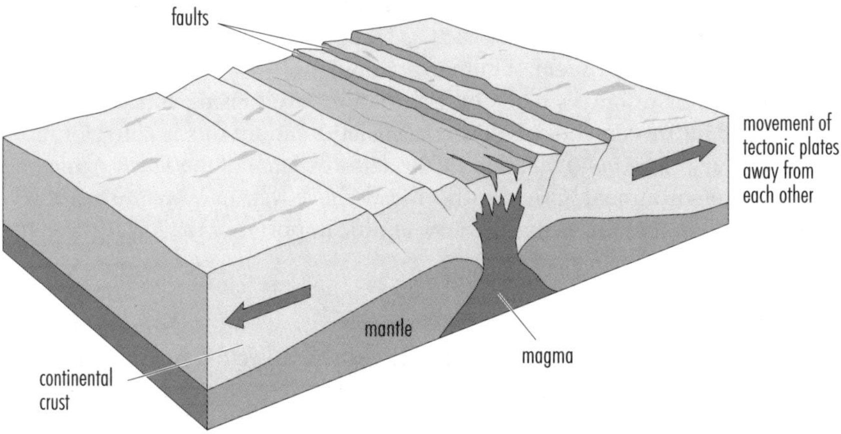

river

A large body of water that flows down a slope along a channel restricted by adjacent banks and levées. A river originates at a point called its *source,* and enters a sea or lake at its *mouth.* Along its length it may be joined by smaller rivers, called *tributaries.* A river and its tributaries are contained within a *drainage basin.* The point at which two rivers join is called the *confluence.*

Rivers are formed and molded over time chiefly by the processes of erosion, and by the transportation and deposition of sediment. Rivers are able to work on the landscape because the energy stored in the water, or potential energy, is converted as it flows downhill into the kinetic energy used for erosion, transportation, and deposition. The amount of potential energy available to a river is proportional to its initial height above sea level. A river follows the path of least resistance downhill, thus minimizing the loss of potential energy to thermal energy (heat) as a result of friction, and maximizing available kinetic energy. Even so, it is estimated that 95 percent of a river's potential energy is used to overcome friction, as it deepens, widens, and lengthens its channel by erosion. This occurs mainly along the channel boundaries, although the internal friction of the water and air resistance on the surface are also important.

One way of classifying rivers is by their stage of development. A youthful stream is typified by a narrow V-shaped valley with numerous waterfalls, lakes, and rapids. Because of the steep gradient of the topography and the river's height above sea level, the rate of erosion is greater than the rate of deposition and downcutting occurs by *vertical corrasion.* These characteristics may also be said to typify a river's *upper course.*

In a mature river, the topography has been eroded down over time and the river's course has a shallow gradient. This mature river is said to be *graded.* Erosion and deposition are delicately balanced as the river meanders (gently curves back and forth) across the extensive floodplain (sometimes called an inner delta). *Horizontal corrasion* is the dominant erosive process. The *floodplain* is an area of periodic flooding along the course of river valleys made up of fine silty material called alluvium deposited by the floodwater. Features of a mature river (or the *lower course* of a river) include extensive meanders, oxbow lakes, and braiding.

Many important floodplains, such as the inner Niger delta in Mali, occur in arid areas where their exceptional fertility has great importance for the local economy. However, using floodplains as the site of towns and villages involves a certain risk, and it is safer to use floodplains for other uses, such as agriculture and parks. Water engineers can predict when flooding is likely and take action to prevent it by studying hydrographs, graphs showing how the discharge of a river varies with time.

Major rivers of the world include the Ganges, the Mississippi, and the Nile, the world's longest river.

river capture

The diversion of the headwaters of one river into a neighboring river. River capture occurs when a stream is carrying out rapid headward erosion (backward erosion at its source). Eventually the stream will cut into the course of a neighboring river, causing the headwaters of that river to be diverted, or *captured.*

The headwaters will then flow down to a lower level (often making a sharp bend, called an *elbow of capture*) over a steep slope, called a *knickpoint.* A waterfall will form here. Rejuvenation then occurs, causing rapid downward erosion.

river cliff

Also *bluff.* A steep slope forming the outer bank of a meander (bend in a river). It is formed by the undercutting of the river current, which flows at its fastest when it sweeps around the outside of the meander.

river terrace

The part of an old floodplain that has been left perched on the side of a river valley. It results from rejuvenation, a renewal in the erosive powers of a river. River terraces are fertile and are often used for farming. They are also commonly chosen as sites for human settlement because they are safe from flooding.

roche moutonnée

An outcrop of tough bedrock having one smooth side and one jagged side, found on the floor of a glacial trough (U-shaped valley). It may be up to 40 m/ 132 ft high. A roche moutonnée is a feature of glacial erosion. As a glacier moved over its surface, ice and debris eroded its upstream side by corrasion, smoothing it and creating long scratches or striations. On the sheltered downstream side fragments of rock were plucked away by the ice, causing it to become steep and jagged.

rock

A constituent of the earth's crust composed of minerals or materials of organic origin that have consolidated into hard masses as igneous, sedimentary, or metamorphic rocks. Rocks are formed from a combination (or aggregate) of minerals, and the property of a rock will depend on its components. Where deposits of economically valuable minerals occur, they are termed *ores.* As a result of weathering, rock breaks down into very small particles that combine with organic materials from plants and animals to form soil. In geology the term *rock* can also include unconsolidated materials such as sand, mud, clay, and peat.

Igneous rock is formed by the cooling and solidification of magma, the molten rock material that originates in the lower part of the earth's crust, or mantle, where it reaches temperatures as high as 1,000°C. The rock may form on or below the earth's surface and is usually crystalline in texture. Larger crystals are more common in rocks such as granite that have cooled slowly within the earth's crust; smaller crystals form in rocks such as basalt that have cooled more rapidly on the surface. Because of their acidic composition, igneous rocks such as granite are particularly susceptible to acid rain.

Sedimentary rocks are formed by the compression of particles deposited by water, wind, or ice. They may be created by the erosion of older rocks, the deposition of organic materials, or they may be formed from chemical precipitates. For example, sandstone is derived from sand particles, limestone from the remains of sea creatures, and gypsum is precipitated from evaporating sea water. Sedimentary rocks are typically deposited in distinct layers or *strata* and many contain fossils.

Metamorphic rocks are formed through the action of high pressure or heat on existing igneous or sedimentary rocks, causing changes to the composition, structure, and texture of the rocks. For example, marble is formed by the effects of heat and pressure on limestone, while granite may be metamorphosed into gneiss, a coarse-grained foliated rock.

Rock studies. The study of the earth's crust and its composition fall under a number of interrelated sciences, each with its own specialists. Among these are geologists, who identify and survey rock formations and determine when and how they were formed; petrologists, who identify and classify the rocks themselves; and mineralogists, who study the mineral contents of the rocks. Paleontologists study the fossil remains of plants and animals found in rocks.

Applications of rock studies. Data from these studies and surveys enable scientists to trace the history of the earth and learn about the kind of life that existed here millions of years ago. The data are also used in locating and mapping deposits of fossil fuels such as coal, oil, and natural gas, and valuable mineral-containing ores providing metals such as aluminum, iron, lead, and tin, and radioactive elements such as radium and uranium. These deposits may lie close to the earth's surface or deep underground, often under oceans. In some regions, entire mountains are composed of deposits of iron or copper ores, while in other regions rocks may contain valuable nonmetallic minerals such as borax and graphite, or precious gems such as diamonds and emeralds.

Rock as construction material and building stone. In addition to the mining and extraction of fuels, metals, minerals, and gems, rocks provide useful building and construction materials. Rock is mined through *quarrying,* and cut into blocks or slabs as building stone, or crushed or broken for other uses in construction work. For instance, cement is made from limestone and, in addition to its use as a bonding material, it can be added to crushed stone, sand, and water to produce strong, durable concrete, which has many applications, such as the construction of roads, runways, and dams.

Among the most widely used building stones are granite, limestone, sandstone, marble, and slate. Granite provides one of the strongest building stones and is resistant to weather, but its hardness makes it difficult to cut and handle. Limestone is a hard and lasting stone that is easily cut and shaped and is widely used for public buildings. The color and texture of the stone can vary with location; for instance, portland stone from the Jurassic rocks of Dorset is white, even textured, and durable, while Bath stone is an oolitic limestone that is honey colored and more porous. Sandstone varies in color and texture; like limestone, it is relatively easy to quarry and work and is used for similar purposes. Marble is a classic stone, worked by both builders and sculptors. Pure marble is white, streaked with veins of black, gray, green, pink, red, and yellow. Slate is fine-grained rock that can be split easily into thin slabs and used as tiles for roofing and flooring. Its color varies from black to green and red.

Rock identification. Rocks can often be identified by their location and appearance. For example, sedimentary rocks lie in stratified, or layered, formations and may contain fossils; many have markings such as old mud cracks or ripple marks caused by waves. Except for volcanic glass, all igneous rocks are solid and crystalline. Some appear dense, with microscopic crystals, and others have larger, easily seen crystals. They occur in volcanic areas, and in intrusive formations that geologists call *batholiths, laccoliths, sills, dikes,* and *stocks.* Many metamorphic rocks have characteristic bands, and are easily split into sheets or slabs. Rock formations and strata are often apparent in the cliffs that line a seashore, or where rivers have gouged out deep channels to form gorges and canyons. They are also revealed when roads are cut through hillsides, or by excavations for quarrying and mining. Rock and fossil collecting has been a popular hobby since the nineteenth century and such sites can provide a treasure trove of finds for the collector.

ruby

The red transparent gem variety of the mineral corundum Al_2O_3, aluminum oxide. Small amounts of chromium oxide, Cr_2O_3, substituting for aluminum oxide, give ruby its color. Natural rubies are found mainly in Myanmar (Burma), but rubies can also be produced artificially and such synthetic stones are used in lasers.

runoff

Water that flows off the land, either through streams or over the surface.

rutile

A titanium oxide mineral, TiO_2, a naturally occurring ore of titanium. It is usually reddish brown to black, with a very bright (adamantine) surface luster. It crystallizes in the tetragonal system. Rutile is common in a wide range of igneous and metamorphic rocks and also occurs concentrated in sands; the coastal sands of eastern and western Australia are a major source. It is also used as a pigment that gives a brilliant white to paint, paper, and plastics.

sabkha

A flat shoreline zone in arid regions above the high-water mark in which the sediments include large amounts of evaporites. These occur in the form of nodules, crusts, and crystalline deposits of halite, anhydrite, and gypsum, as well as mineral grains of various sorts. Some of the evaporites form from rapid evaporation of marine waters soaking through from the bordering tidal flats, but some can be derived also from sediment-laden continental waters coming down from adjoining highlands. Sabkhas are common features along the Persian Gulf. Former sabkha environments can be identified from the sedimentary record of much of the oil-rich Permian Basin of Texas, as an ancient sea moved back and forth across the continent.

Sahel

Arabic *sahil,* meaning *coast.* A marginal area to the south of the Sahara, from Senegal to Somalia, which experiences desertlike conditions during periods of low rainfall. The desertification is partly due to climatic fluctuations but has also been caused by the pressures of a rapidly expanding population, which has led to overgrazing and the destruction of trees and scrub for fuel. In recent years many famines have taken place in the area. The average rainfall in the Sahel ranges from 100 mm/4 in to 500 mm/20 in per year, but the rainfall over the past 30 years has been significantly below average. The resulting famine and disease are further aggravated by civil wars. The areas most affected are Ethiopia and the Sudan.

salinization

The accumulation of salt in water or soil; it is a factor in desertification.

salt, common

Sodium chloride, NaCl, a white crystalline solid found dissolved in sea water and as rock salt (the mineral halite) in large deposits and salt domes. Common salt is used extensively in the food industry as a preservative and for flavoring, and in the chemical industry in the making of chlorine and sodium. Although common salt is an essential part of our diet, some medical experts believe that excess salt can lead to high blood pressure and increased risk of heart attacks. Salt has historically been considered a sustaining substance, often taking on religious significance in ancient cultures. Roman soldiers were paid part of their wages as salt allowance (Latin *salerium argentinium*), hence the term *salary.*

salt marsh

A wetland with halophytic vegetation (tolerant to sea water). Salt marshes develop around estuaries and on the sheltered side of sand and gravel spits. Salt marshes usually have a network of creeks and drainage channels by which tidal waters enter and leave the marsh.

saltation

The bouncing of rock particles along a river bed. It is the means by which bedload (material that is too heavy to be carried in suspension) is transported downstream. The term is also used to describe the movement of sand particles bounced along by the wind.

San Andreas fault

A fault stretching for 1,125 km/700 mi northwest to southeast through the state of California. It marks a conservative plate margin, where two plates slide past each other (*see* **plate tectonics**). Friction is created as the coastal Pacific plate moves to the northwest, rubbing against the American continental plate, which is moving slowly southeast. The relative movement is only about 5 cm/2 in a year, which means that Los Angeles will reach San Francisco's latitude in 10 million years. The friction caused by the tectonic movement gives rise to frequent, destructive earthquakes. For example, in 1906 an earthquake originating from the fault almost destroyed San Francisco and killed about 700 people. The San Andreas fault has mountains on either side, uplifted by the movements of the plates; rivers flow along its course because it represents a line of weakness that can be easily eroded.

sand

Loose grains of rock, sized 0.0625–2.00 mm/0.0025–0.08 in in diameter, consisting most commonly of quartz, but owing their varying color to mixtures of other minerals. Sand is used in cement making, as an abrasive, in glass making, and for other purposes. Sands are classified into marine, freshwater, glacial, and terrestrial. Some "light" soils contain up to 50 percent sand. Sands may eventually consolidate into sandstone.

sandbar

A ridge of sand built up by the currents across the mouth of a river or bay. A sandbar may be entirely underwater or it may form an elongated island that breaks the surface. A sandbar stretching out from a headland is a *sand spit.* Coastal bars can extend across estuaries to form *bay bars.*

sandstone

Sedimentary rocks formed from the consolidation of sand, with sand-sized grains (0.0625–2 mm/0.0025–0.08 in) in a matrix or cement. Their principal component is quartz. Sandstones are commonly permeable and porous, and may form freshwater aquifers. They are mainly used as building materials. Sandstones are classified according to the matrix or cement material (whether derived from clay or silt; for example, as calcareous sandstone, ferruginous sandstone, or siliceous sandstone).

sapphire

A deep blue, transparent gem variety of the mineral corundum Al_2O_3, aluminum oxide. Small amounts of iron and titanium give it its color. A corundum gem of any color except red (which is a ruby) can be called a sapphire; for example, yellow sapphire.

sard

A yellow or red-brown variety of chalcedony.

savanna

Also *savannah.* Extensive open tropical grasslands, with scattered trees and shrubs. Savannas cover large areas of Africa, North and South America, and northern Australia. The soil is acidic and sandy and generally considered suitable only as pasture for low-density grazing. The name was originally given by Spaniards to the treeless plains of the tropical South American prairies. Most of North America's savannas have been built over. A new strain of rice suitable for savanna conditions was developed 1992. It not only grew successfully under test conditions in Colombia but also improved pasture quality so grazing numbers could be increased twentyfold.

scapolite

A group of white or grayish minerals, silicates of sodium, aluminum, and calcium, common in metamorphosed limestones and forming at high temperatures and pressures.

scarp and dip

The two slopes formed when a sedimentary bed outcrops as a landscape feature. The scarp is the slope that cuts across the bedding plane; the dip is the opposite slope which follows the bedding plane. The scarp is usually steep, while the dip is a gentle slope.

schist

A metamorphic rock containing mica or another platy or elongate mineral the crystals of which are aligned to give a foliation (planar texture) known as *schistosity.* Schist may contain additional minerals such as garnet.

scree
A pile of rubble and sediment that collects at the foot of a mountain range or cliff. The rock fragments that form scree are usually broken off by the action of frost (freeze-thaw weathering). With time, the rock waste builds up into a heap or sheet of rubble that may eventually bury even the upper cliffs. The growth of the scree then stops. Usually, however, erosional forces remove the rock waste so that the scree stays restricted to lower slopes.

sea breeze
A gentle coastal wind blowing off the sea toward the land. It is most noticeable in summer when the warm land surface heats the air above it and causes it to rise. Cooler air from the sea is drawn in to replace the rising air, so causing an onshore breeze. At night and in winter, air may move in the opposite direction, forming a land breeze.

sea level
The average height of the surface of the oceans and seas measured throughout the tidal cycle at hourly intervals and computed over a 19-year period. It is used as a datum plane from which elevations and depths are measured. Factors affecting sea level are (1) a combination of naturally high tides and storm surges, as sometimes happens along the low-lying coasts of Germany and the Netherlands; (2) the water walls created by typhoons and hurricanes, such as often hit Bangladesh; (3) underwater upheavals in Earth's crust which may cause a tsunami; and (4) global temperature change melting polar ice caps.

sea-floor spreading
The growth of the ocean crust outward (sideways) from ocean ridges. The concept of sea-floor spreading has been combined with that of continental drift and incorporated into plate tectonics. Sea-floor spreading was proposed in 1960 by U.S. geologist Harry Hess (1906–1969), based on his observations of ocean ridges and the relative youth of all ocean beds. In 1963 British geophysicists Fred Vine and Drummond Matthews observed that the floor of the Atlantic Ocean was made up of rocks that could be arranged in strips, each strip being magnetized either normally or reversely (due to changes in the earth's polarity when the North Pole becomes the South Pole and vice versa, termed *polar reversal*). These strips were parallel and formed identical patterns on both sides of the ocean ridge. The implication was that each strip was formed at some stage in geological time when the magnetic field was polarized in a certain way. The seafloor magnetic-reversal patterns could be matched to dated magnetic reversals found in terrestrial rock. It could then be shown that new rock forms continuously and spreads away from the ocean ridges, with the oldest rock located farthest away from the midline. The observation was made independently in 1963 by Canadian geologist Lawrence Morley, studying an ocean ridge in the Pacific near Vancouver Island. Confirmation came when sediments were discovered to be deeper further away from the oceanic ridge, because the rock there had been in existence longer and had had more time to accumulate sediment.

season

A period of the year having a characteristic climate. The change in seasons is mainly due to the change in attitude of the earth's axis in relation to the sun, and hence the position of the sun in the sky at a particular place. In temperate latitudes four seasons are recognized: spring, summer, autumn (fall), and winter. Tropical regions have two seasons—the wet and the dry. Monsoon areas around the Indian Ocean have three seasons: the cold, the hot, and the rainy.

The northern temperate latitudes have summer when the southern temperate latitudes have winter, and vice versa. During winter, the sun is low in the sky and has less heating effect because of the oblique angle of incidence and because the sunlight has further to travel through the atmosphere. The differences between the seasons are more marked inland than near the coast, where the sea has a moderating effect on temperatures. In polar regions the change between summer and winter is abrupt; spring and autumn are hardly perceivable. In tropical regions, the belt of rain associated with the trade winds moves north and south with the sun, as do the dry conditions associated with the belts of high pressure near the Tropics. The monsoon's three seasons result from the influence of the Indian Ocean on the surrounding landmass of Asia in that area.

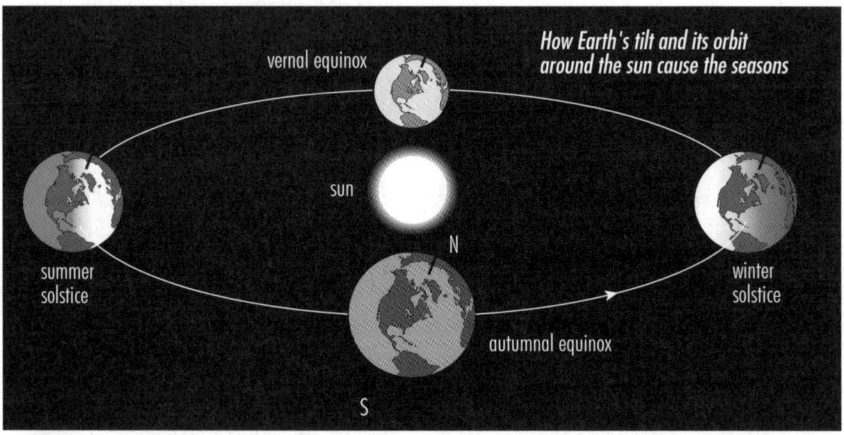

secular variation

Changes in the position of Earth's magnetic poles measured with respect to geographical positions, such as the North Pole, throughout geological time.

sediment

Any loose material that has "settled" after being deposited from suspension in water, ice, or air, generally as the water current or wind speed decreases. Typical sediments are, in order of increasing coarseness, clay, mud, silt, sand, gravel, pebbles, cobbles, and boulders.

Sediments differ from sedimentary rocks in which deposits are fused together in a solid mass of rock by a process called *lithification.* Pebbles are cemented into conglomerates; sands become sandstones; muds become mudstones or shales; peat is transformed into coal.

sedimentary rock

A rock formed by the accumulation and cementation of deposits that have been laid down by water, wind, ice, or gravity. Sedimentary rocks cover more than two-thirds of the earth's surface and comprise three major categories: clastic, chemically precipitated, and organic (or biogenic). Clastic sediments are the largest group and are composed of fragments of preexisting rocks; they include clays, sands, and gravels. Chemical precipitates include some limestones and evaporated deposits such as gypsum and halite (rock salt). Coal, oil shale, and limestone made of fossil material are examples of organic sedimentary rocks.

Most sedimentary rocks show distinct layering (stratification), caused by alterations in composition or by changes in rock type. These strata may become folded or fractured by the movement of the earth's crust, a process known as *deformation.*

seiche

A pendulous movement seen in large areas of water resembling a tide. It was originally observed on Lake Geneva and is created either by earthquakes, wind, or other atmospheric phenomena.

seismic gap theory

A theory that aims to predict the location of earthquakes. When records of past earthquakes are studied and plotted onto a map, it becomes possible to identify areas along a fault or plate margin where an earthquake should be due. Such areas are called *seismic gaps.* According to the theory, an area that has not had an earthquake for some time will have a great deal of stress building up, which must eventually be released in the form of an earthquake.

Although the seismic gap theory can suggest areas that are likely to experience an earthquake, it does not enable scientists to predict when that earthquake will occur. Research carried out in the vicinity of the San Andreas fault in California has identified a seismic gap at the town of Parkfield, near San Francisco. Only time will tell whether the prediction will prove to be correct.

seismic wave

An energy wave generated by an earthquake or an artificial explosion. There are two types of seismic waves: *body waves* that travel through the earth's interior, and *surface waves* that travel through the surface layers of the crust and can be felt as the shaking of the ground, as in an earthquake.

Body waves. There are two types of body waves: *P-waves* and *S-waves,* so named because they are the primary and secondary waves detected by a seismograph. *P-waves* are longitudinal waves (wave motion in the direction the wave is traveling), the compressions and rarefactions of which resemble those of a sound wave. *S-waves* are transverse waves or shear waves, involving a back-and-forth shearing motion at right angles to the direction the wave is traveling (*see* **wave**).

Because liquids have no resistance to shear and cannot sustain a shear wave, S-waves cannot travel through liquid material. The earth's outer core is believed to be liquid because S-waves disappear at the mantle-core boundary, while P-waves do not.

Surface waves. Surface waves travel in the surface and subsurface layers of the crust. *Rayleigh waves* travel along the free surface (the uppermost layer) of a solid material. The motion of particles is elliptical, like a water wave, creating the rolling motion often felt during an earthquake. The Raleigh wave is named for Lord Raleigh, an English physicist who predicted it. *Love waves* (named for English mathematician A. E. H. Love) are transverse waves trapped in a subsurface layer due to different densities in the rock layers above and below. They have a horizontal side-to-side shaking motion transverse (at right angles) to the direction the wave is traveling.

seismogram

Also *seismic record.* A trace, or graph, of an earthquake's activity over time, recorded by a seismograph. It is used to determine the magnitude and duration of an earthquake.

seismograph

An instrument used to record the activity of an earthquake. A heavy inert weight is suspended by a spring and attached to this is a pen that is in contact with paper on a rotating drum. During an earthquake the instrument frame and drum move, causing the pen to record a zigzag line on the paper; the pen does not move.

seismology

The study of earthquakes and how their shock waves travel through the earth. By examining the global pattern of waves produced by an earthquake, seismologists can deduce the nature of the materials through which they have passed. This leads to an understanding of the earth's internal structure.

On a smaller scale, artificial earthquake waves, generated by explosions or mechanical vibrators, can be used to search for subsurface features in, for example, oil or mineral exploration. Earthquake waves from underground nuclear explosions can be distinguished from natural waves by their shorter wavelength and higher frequency.

selva

An equatorial rainforest, such as that in the Amazon basin in South America.

serpentine

A group of minerals, hydrous magnesium silicate, $Mg_3Si_2O_5(OH)_4$, occurring in soft metamorphic rocks and usually dark green. The fibrous form chrysotile is a source of asbestos; other forms are antigorite and lizardite. Serpentine minerals are formed by hydration of ultramafic rocks during metamorphism. Rare snake-patterned forms are used in ornamental carving.

shale

A fine-grained and finely layered sedimentary rock composed of silt and clay. It is a weak rock, splitting easily along bedding planes to form thin, even slabs (by contrast, mudstone splits into irregular flakes). Oil shale contains kerogen, a solid bituminous material that yields petroleum when heated.

shelf sea

A relatively shallow sea, usually no deeper than 200 m/650 ft, overlying the continental shelf around the coastlines. Most fishing and marine mineral exploitations are carried out in shelf seas.

shield

An alternative name for *craton,* the ancient core of a continent.

shield volcano

A broad flat volcano formed at a constructive margin between tectonic plates or over a hot spot. The magma (molten rock) associated with shield volcanoes is usually basalt, thin and free flowing. An example is Mauna Loa in Hawaii. A composite volcano, on the other hand, is formed at a destructive margin.

sial

In geochemistry and geophysics, the substance of the earth's continental crust, as distinct from the sima of the ocean crust. The name, now used rarely, is derived from silica and alumina, its two main chemical constituents. Sial is often rich in granite.

silica

Silicon dioxide, SiO_2. The most familiar mineral form is quartz, the principle constituent of white beach sand. Other forms of silica include chalcedony, chert, opal, tridymite, coesite, and cristobalite.

silicate

A group of minerals containing silicon and oxygen in tetrahedral units of SiO_4, bound together in various ways to form specific structural types. Silicates are the chief rock-forming minerals. Most rocks are composed, wholly or in part, of silicates (the main exception being limestones). Glass is a manufactured complex polysilicate material in which other elements (boron in borosilicate glass) have been incorporated.

Generally, additional cations are present in the structure, especially Al^{3+}, Fe^{2+}, Mg^{2+}, Ca^{2+}, Na^{+}, K^{+}, but quartz and other polymorphs of SiO_2 are also considered to be silicates; stishovite (a high-pressure form of SiO_2) is a rare exception to the usual tetrahedral coordination of silica and oxygen.

In *orthosilicates,* the oxygens bonded to silicon are not shared between SiO_4 tetrahedra. All other silicate structures involve some degree of oxygen sharing between adjacent tetrahedra. For example, beryl is a *ring silicate* based on tetrahedra linked by sharing oxygens to form a circle. Pyroxenes are single *chain silicates,* with chains of linked tetrahedra extending in one direction through the structure; amphiboles are similar but have double chains of tetrahedra. In micas, which are *sheet silicates,* the tetrahedra are joined to form continuous sheets that are stacked upon one another. *Framework silicates,* such as feldspars and quartz, are based on three-dimensional frameworks of tetrahedra in which all oxygen molecules of the SiO_4 tetrahedra are shared.

sill

A sheet of igneous rock created by the intrusion of magma (molten rock) between layers of preexisting rock. (A dike, by contrast, is formed when magma cuts *across* layers of rock.) The Palisades Sill along the Hudson River in New York is an example. A sill is usually formed of diabase, a rock that is extremely resistant to erosion and weathering, and often forms ridges in the landscape or cuts across rivers to create waterfalls.

sillimanite

Aluminum silicate, Al_2SiO_5, a mineral that occurs either as white to brownish prismatic crystals or as minute white fibers. It is an indicator of high temperature conditions in metamorphic rocks formed from clay sediments. Andalusite, kyanite, and sillimanite are all polymorphs of Al_2SiO_5.

silt

Sediment intermediate in coarseness between clay and sand; its grains have a diameter of 0.002–0.02 mm/0.00008–0.0008 in. Silt is usually deposited in rivers, and so the term is often used generically to mean a river deposit, as in the silting up of a channel.

Silurian

The period of geological time 439 million to 409 million years ago, the third period of the Paleozoic era. Silurian sediments are mostly marine and consist of shales and limestone. Luxuriant reefs were built by coral-like organisms. The first land plants began to evolve during this period, and there were many ostracoderms (armored jawless fishes). The first jawed fishes (called *acanthodians*) also appeared.

sima

In geochemistry and geophysics, the substance of the earth's oceanic crust, as distinct from the sial of the continental crust. The name, now used rarely, is derived from silica and magnesia, its two main chemical constituents.

sink hole

A funnel-shaped hollow in an area of limestone. A sink hole is usually formed by the enlargement of a joint, or crack, by carbonation (the dissolving effect of water). It should not be confused with a swallow hole, or *swallet,* which is the opening through which a stream disappears underground when it passes onto limestone.

sirocco

A hot, normally dry and dust-laden wind that blows from the deserts of northern Africa across the Mediterranean into southern Europe. It occurs mainly in the spring. The name *sirocco* is also applied to any hot oppressive wind.

slate

A fine-grained, usually gray metamorphic rock that splits readily into thin slabs along its cleavage planes. It is the metamorphic equivalent of shale. Slate is highly resistant to atmospheric conditions and can be used for writing on with chalk (actually gypsum). Quarrying slate takes such skill and time that it is now seldom used for roof and sill material except in restoring historic buildings.

sleet

Precipitation consisting of a mixture of water and ice. The term often describes frozen raindrops or snow that has melted and refrozen.

slip-off slope

A gentle slope forming the inner bank of a meander (bend in a river). It is formed by the deposition of fine silt, or alluvium, by slow-flowing water. As water passes around a meander, the fastest current sweeps past the outer bank, eroding it to form a steep river cliff. Water flows more slowly past the inner bank, and as it reduces speed the material it carries is deposited around the bank to form a slip-off slope.

smog

A natural fog containing impurities, mainly nitrogen oxides (NO_x) and volatile organic compounds (VOCs) from domestic fires, industrial furnaces, certain power stations, and internal-combustion engines (gasoline or diesel). It can cause substantial illness and loss of life, particularly among chronic bronchitics, and damage to wildlife.

Photochemical smog is mainly prevalent in the summer as it is caused by chemical reaction between strong sunlight and vehicle exhaust fumes. Such smogs create a buildup of ozone and nitrogen oxides that cause adverse symptoms, including coughing and eye irritation, and in extreme cases can kill. The use of smokeless fuels, the treatment of effluent, and penalties for excessive smoke from poorly maintained and operated vehicles can be effective in reducing smog but it still occurs in many cities throughout the world.

smoker

Also *hydrothermal vent.* A crack in the ocean floor, associated with an ocean ridge, through which hot mineral-rich groundwater erupts into the sea, forming thick clouds of suspended material. The clouds may be dark or light, depending on the mineral content, thus producing *white smokers* or *black smokers*. Seawater percolating through the sediments and crust is heated in the active area beneath and dissolves minerals from the hot rocks. As the charged water is returned to the ocean, the sudden cooling causes these minerals to precipitate from solution, so forming the suspension. The chemical-rich water around a smoker gives rise to colonies of bacteria, and these form the basis of food chains that can be sustained without sunlight and photosynthesis. Strange animals that live in such regions include huge tube worms 2 m/6 ft long, giant clams, and species of crab, anemone, and shrimp found nowhere else.

snout

The front end of a glacier, representing the furthest advance of the ice at any one time. Deep cracks, or crevasses, and ice falls are common. Because the snout is the lowest point of a glacier, it tends to be affected by the warmest weather. Considerable melting takes place, and so it is here that much of the rocky material carried by the glacier becomes deposited. Material dumped just ahead of the snout may form a terminal moraine. The advance or retreat of the snout depends upon the glacier budget—the balance between accumulation (the addition of snow and ice to the glacier) and ablation (their loss by melting and evaporation).

snow

Precipitation in the form of soft, white crystalline flakes caused by the condensation in air of excess water vapor below freezing point. Light reflecting in the crystals, which have a basic hexagonal (six-sided) geometry, gives snow its white appearance.

soapstone

A compact, massive form of impure talc.

soil

A loose covering of broken rocky material and decaying organic matter overlying the bedrock of the earth's surface. It is comprised of minerals, organic matter (called *humus*) derived from decomposed plants and organisms, living organisms, air, and water. Soils differ according to climate, parent material, rainfall, relief of the bedrock, and their proportion of organic material. The study of soils is called *pedology*.

A soil can be described in terms of its *soil profile,* that is, a vertical cross-section from ground level to the bedrock on which the soils sits. The profile is divided into layers called *horizons*. The A horizon, or *topsoil,* is the uppermost layer, consisting primarily of humus and living organisms and some mineral material. Most soluble material has been leached from this layer or washed down to the B horizon. The B horizon, or *subsoil,* is the layer where most of the nutrients accumulate and is enriched in clay minerals. The C horizon is the layer of weathered parent material at the base of the soil.

Two common soils are the *podzol* and the *chernozem* soil. The podzol is common in coniferous forest regions where precipitation exceeds evaporation. The A horizon consists of a very thin litter of organic material producing a poor humus. Needles take a long time to decompose. The relatively heavy precipitation causes leaching of minerals, as nutrients are washed downward. Chernozem soils are found in grassland regions, where evaporation exceeds precipitation. The A horizon is rich in humus due to decomposition of a thick litter of dead grass at the surface. Minerals and moisture migrate upward due to evaporation, leaving the B and A horizons enriched. The organic content of soil is widely variable, ranging from zero in some desert soils to almost 100 percent in peats.

Soils influence the type of agriculture employed in a particular region. Light well-drained soils favor arable farming, whereas heavy clay soils give rise to lush pastureland.

soil creep

The gradual movement of soil down a slope in response to gravity. This eventually results in a mass downward movement of soil on the slope. Evidence of soil creep includes the formation of terracettes (steplike ridges along the hillside), leaning walls and telegraph poles, and trees that grow in a curve to counteract progressive leaning.

soil depletion

A decrease in soil quality over time. Causes include loss of nutrients from overfarming, erosion by wind, and chemical imbalances caused by acid rain.

soil erosion

The wearing away and redistribution of the earth's soil layer. It is caused by the action of water, wind, and ice, and also by improper methods of agriculture. If unchecked, soil erosion results in the formation of deserts (desertification). It has been estimated that 20 percent of the world's cultivated topsoil was lost between 1950 and 1990. If the rate of erosion exceeds the rate of soil formation (from rock and decomposing organic matter), the land will become infertile. The removal of forests (deforestation) or other vegetation often leads to serious soil erosion, because plant roots bind soil, and without them the soil is free to wash or blow away, as in the American dust bowl. The effect is worse on hillsides, and there has been devastating loss of soil where forests have been cleared from mountainsides, as in Madagascar.

Improved agricultural practices such as contour plowing are needed to combat soil erosion. Windbreaks, such as hedges or strips planted with coarse grass, are valuable, and organic farming can reduce soil erosion by as much as 75 percent.

Soil degradation and erosion are becoming as serious as the loss of the rainforest. It is estimated that more than 10 percent of the world's soil lost a large amount of its natural fertility during the latter half of the twentieth century. Some of the worst losses are in Europe, where 17 percent of the soil is damaged by human activity, such as mechanized farming and fallout from acid rain. Mexico and Central America have 24 percent of their soil highly degraded, mostly as a result of deforestation.

solar pond

A natural or artificial "pond," such as the Dead Sea, in which salt becomes more soluble in the sun's heat. Water at the bottom becomes saltier and hotter, and is insulated by the less salty water layer at the top. Temperatures at the bottom reach about 100°C/212°F and can be used to generate electricity.

solifluction

The downhill movement of topsoil that has become saturated with water. Solifluction is common in periglacial environments (those bordering glacial areas) during the summer months, when the frozen topsoil melts to form an unstable soggy mass. This may then flow slowly downhill under gravity to form a *solifluction lobe* (a tonguelike feature). Solifluction material, or *head,* is found at the bottom of chalk valleys in southern England; it is partly responsible for the rolling landscape typical of chalk scenery.

solution

The process by which the minerals in a rock are dissolved in water. It is also referred to as *corrosion.* Solution is one of the processes of erosion, as well as weathering (in which the dissolution of rock occurs without transportation of the dissolved material). An example of this is when weakly acidic rainfall causes carbonation.

Solution commonly affects limestone and chalk, both forms of calcium carbonate. It can occur in coastal environments along with corrasion and hydraulic action, producing features like the white cliffs of Dover, as well as fluvial (river) environments. In groundwater environments of predominantly limestone solution produces karst topography, forming features such as sink holes, caves, and limestone pavement. Solution also is responsible for the weathering of buildings, monuments, and other man-made structures.

Southern Ocean

A corridor linking the Pacific, Atlantic, and Indian Oceans, all of which receive cold polar water from the world's largest ocean surface current, the Antarctic Circumpolar Current, which passes through the Southern Ocean.

speleology

The scientific study of caves, their origin, development, physical structure, flora, fauna, folklore, exploration, mapping, photography, cave diving, and rescue work. *Potholing,* which involves following the course of underground rivers or streams, has become a popular sport. Speleology first developed in France in the late nineteenth century, where the Société de Spéléologie was founded in 1895.

sphalerite

A mineral composed of zinc sulfide with a small proportion of iron, (Zn,Fe)S. It is the chief ore of zinc. Sphalerite is brown with a nonmetallic luster unless an appreciable amount of iron is present (up to 26 percent by weight). Sphalerite usually occurs in ore veins in limestones, where it is often associated with galena. It crystallizes in the cubic system but does not normally form perfect cubes.

spinel

Any of a group of "mixed oxide" minerals consisting mainly of the oxides of magnesium and aluminum, $MgAl_2O_4$ and $FeAl_2O_4$. Spinels crystallize in the cubic system, forming octahedral crystals. They are found in high-temperature igneous and metamorphic rocks. The aluminum oxide spinel contains gem varieties, such as the ruby spinels of Sri Lanka and Myanmar (Burma).

spit

A ridge of sand or gravel projecting from the land into a body of water. It is formed by the interruption of longshore drift due to wave interaction with tides, currents, or a bend in the coastline. The consequent decrease in wave energy causes more material to be deposited than is transported down the coast, building up a finger of sand that points in the direction of the longshore drift. Deposition in the brackish water behind a spit may result in the formation of a salt marsh.

spring

A natural flow of water from the ground, formed at the point of intersection of the water table and the ground's surface. The source of water is rain that has percolated through the overlying rocks. During its underground passage, the water may have dissolved mineral substances that may then be precipitated at the spring (hence, a mineral spring). A spring may be continuous or intermittent, and depends on the position of the water table and the topography (surface features).

spring line

A geological feature where water springs up in several places along the edge of a permeable rock escarpment. Springline settlements may become established around this.

sprite

A rare thunderstorm-related luminous flash. Sprites occur in the mesosphere, at altitudes of 50–90 km/31–56 mi. They are electrical, like lightning, and arise when the electrical field that occurs between the thundercloud top and the ionosphere (ionized layer of the earth's atmosphere) draws electrons upward from the cloud. If the air is thin and this field is strong the electrons accelerate rapidly, transferring kinetic energy to molecules they collide with. The excited molecules then discharge this energy as a light flash, which brightens the sky as opposed to a "bolt" of lightning.

spur

A ridge of rock jutting out into a valley or plain. In mountainous areas, rivers often flow around interlocking spurs because they are not powerful enough to erode through the spurs. Spurs may be eroded away by large and powerful glaciers to form truncated spurs.

stack

An isolated pillar of rock that has become separated from a headland by coastal erosion. It is usually formed by the collapse of an arch. Further erosion will reduce it to a stump, which is exposed only at low tide. Good examples of stacks can be found in the chalk cliffs off the coast of Normandy and the Needles, off the Isle of Wight, which are also formed of chalk.

stalactite and stalagmite

Cave structures formed by the deposition of calcite dissolved in ground water. *Stalactites* grow downward from the roofs or walls and can be icicle shaped, straw shaped, curtain shaped, or formed as terraces. *Stalagmites* grow upward from the cave floor and can be conical, fir cone shaped, or resemble a stack of saucers. Growing stalactites and stalagmites may meet to form a continuous column from floor to ceiling.

Stalactites are formed when groundwater, hanging as a drip, loses a proportion of its carbon dioxide into the air of the cave. This reduces the amount of calcite that can be held in solution, and a small trace of calcite is deposited. Successive drips build up the stalactite over many years. In stalagmite formation the calcite comes out of the solution because of agitation—the shock of a drop of water hitting the floor is sufficient to remove some calcite from the drop. The different shapes result from the splashing of the falling water.

staurolite

A silicate mineral, $(Fe,Mg)_2(Al,Fe)_9Si_4O_{20}(OH)_2$. It forms brown crystals that may be twined in the form of a cross. It is a useful indicator of medium-grade (moderate temperature and pressure) metamorphism in metamorphic rocks formed from clay sediments.

steppe

The temperate grasslands of Europe and Asia. Sometimes the term refers to other temperate grasslands and semiarid desert edges.

Stevenson screen

A box designed to house weather-measuring instruments such as thermometers. It is kept off the ground by legs, has louvered sides to encourage the free passage of air, and is painted white to reflect heat radiation, because what is measured is the temperature of the air, not of the sunshine.

stishovite

The highest density, highest pressure polymorph (substance with the same chemical composition but different crystal structure) of silica, SiO_2, in which the crystal structure is the same as that of the mineral rutile, titanium dioxide. Rare in nature, it is found primarily near meteor impact craters. It is thought to be a rare constituent of the lower mantle.

storm surge

An abnormally high tide brought about by a combination of a deep atmospheric depression (very low pressure) over a shallow sea area, high spring tides, and winds blowing from the appropriate direction. A storm surge can cause severe flooding of lowland coastal regions and river estuaries.

Bangladesh is particularly prone to surges, being sited on a low-lying delta where the Indian Ocean funnels into the Bay of Bengal. In May 1991 125,000 people were killed there in such a disaster. In February 1953 more than 2,000 people died when a North Sea surge struck the Dutch and English coasts.

strata

Singular *stratum.* Layers or beds of sedimentary rock.

stratigraphy

The branch of geology that deals with the sequence of formation of sedimentary rock layers and the conditions under which they were formed. Its basis was developed by William Smith, a British canal engineer. The basic principle of superimposition establishes that upper layers or deposits have accumulated later in time than the lower ones.

Stratigraphy involves both the investigation of sedimentary structures to determine past environments represented by rocks, and the study of fossils for identifying and dating particular beds of rock. A body of rock strata with a set of unifying characteristics indicative of an environment is called a *facies.*

Stratigraphic units can be grouped in terms of time or lithology (rock type). Strata that were deposited at the same time belong to a single *chronostratigraphic unit* but need not be the same lithology. Strata of a specific lithology can be grouped into a *lithostratigraphic unit* but are not necessarily the same age.

Stratigraphy in the interpretation of archeological excavations provides a relative chronology for the levels and the artifacts within rock beds. It is the principal means by which the context of archeological deposits is evaluated.

stratosphere

That part of the atmosphere 10–40 km/6–25 mi from the earth's surface, where the temperature slowly rises from a low of -55°C/-67°F to around 0°C/32°F. The air is rarefied and at around 25 km/15 mi much ozone is concentrated.

stratovolcano

Another term for *composite volcano,* a type of explosive volcano made up of alternate ash and lava layers.

striation

A scratch inscribed in a rock by the movement of a glacier or fault. Glacial striations are caused by the scraping of rocky debris embedded in the base of the glacier (corrasion). Striations provide a useful indicator of the direction of ice or fault motion. They are common features of roche moutonnées.

strike

The compass direction of a horizontal line on a planar structural surface, such as a fault plane, bedding plane, or the trend of a structural feature, such as the axis of a fold. Strike is 90° from dip.

strike-slip fault

The common name for a lateral fault in which the motion is sideways in the direction of the strike of the fault.

stromatolite

A mound produced in shallow water by mats of algae that trap mud particles. Another mat grows on the trapped mud layer and this traps another layer of mud and so on. The stromatolite grows to heights of a meter or so. They are uncommon today, but their fossils are among the earliest evidence for living things—over 2 billion years old.

stump

A low outcrop of rock formed by the erosion of a coastal stack. Unlike a stack, which is exposed at all times, a stump is exposed only at low tide. Eventually it will be worn away completely, leaving a wave-cut platform.

subduction zone

A region where two plates of the earth's rigid lithosphere collide and one plate descends below the other into the weaker asthenosphere. Subduction occurs along ocean trenches, most of which encircle the Pacific Ocean; portions of the ocean plate slide beneath other plates carrying continents.

Ocean trenches are usually associated with volcanic island arcs and deep-focus earthquakes (more than 300 km/185 mi below the surface), both the result of disturbances caused by the plate subduction. The Aleutian Trench bordering Alaska is an example of an active subduction zone, which has produced the Aleutian Island arc.

subglacial

Of or pertaining to conditions beneath a glacier. Subglacial rivers are those that flow under a glacier; subglacial material is debris that has been deposited beneath glacier ice. Features formed subglacially include drumlins and eskers.

sunshine recorder

A device for recording the hours of sunlight during a day. The Campbell-Stokes sunshine recorder consists of a glass sphere that focuses the sun's rays on a graduated paper strip. A track is burned along the strip corresponding to the time that the sun is shining.

supraglacial

Of or pertaining to conditions on top of a glacier. A supraglacial stream flows over the surface of the glacier; supraglacial material collected on top of a glacier may be deposited to form lateral and medial moraines.

surface runoff
The overland transfer of water after a rainfall. It is the most rapid way in which water reaches a river. The amount of surface runoff increases given (1) heavy and prolonged rainfall, (2) steep gradients, (3) lack of vegetation cover, and (4) saturated or frozen soil. A hydrograph can indicate the time the runoff takes to reach the river. Throughflow is another way water reaches a river.

surge
An abnormally high tide; *see* **storm surge.**

surging glacier
A glacier that has periods, generally lasting one to four years, of very rapid flow (up to several meters per hour compared to normal glaciers, which move at a rate of several meters per year) followed by periods of stagnation lasting up to ten years. Surging glaciers are heavily crevassed.

surveying
The accurate measuring of the earth's crust, or of land features or buildings. It is used to establish boundaries, and to evaluate the topography for engineering work. The measurements used are both linear and angular, and geometry and trigonometry are applied in the calculations.

swallet
An alternative name for a *swallow hole.*

swallow hole
Also *swallet.* A hole, often found in limestone areas, through which a surface stream disappears underground. It usually leads to an underground network of caves.

swamp
A region of low-lying land that is permanently saturated with water and usually overgrown with vegetation; for example, the Everglades of Florida. A swamp often occurs where a lake has filled up with sediment and plant material. The flat surface so formed means that runoff is slow, and the water table is always close to the surface. The high humus content of swamp soil means that good agricultural soil can be obtained by draining.

swash
The advance of water and sediment up a beach as a wave breaks. Swash plays a significant role in the movement of beach material by longshore drift, and is responsible for throwing gravel and pebbles up a beach to create ridges called *berms.*

syenite
A gray, crystalline, plutonic (intrusive) igneous rock, consisting of feldspar and hornblende; other minerals may also be present, including small amounts of quartz.

syncline

A geological term for a fold in the rocks of the earth's crust in which the layers or beds dip inward, thus forming a troughlike structure with a sag in the middle. The center of the syncline contains stratigraphically younger rocks. The opposite structure, with the beds arching upward, is an anticline.

synoptic chart

A weather chart in which symbols are used to represent the weather conditions experienced over an area at a particular time. Synoptic charts appear on television and newspaper forecasts, although the symbols used may differ.

taiga

Also *boreal forest.* The Russian name for the forest zone south of the tundra, found across the Northern Hemisphere. Here, dense forests of conifers (spruces and hemlocks), birches, and poplars occupy glaciated regions punctuated with cold lakes, streams, bogs, and marshes. Winters are prolonged and very cold, but the summer is warm enough to promote dense growth. The varied fauna and flora are in delicate balance because the conditions of life are so precarious. This ecology is threatened by mining, forestry, and pipeline construction.

talc

A mineral, hydrous magnesium silicate, $Mg_3Si_4O_{10}(OH)_2$. It occurs in tabular crystals, but the massive impure form, known as *steatite* or *soapstone,* is more common. It is formed by the alteration of magnesium compounds and is usually found in metamorphic rocks. Talc is very soft, ranked 1 on the Mohs scale of hardness. It is used in powdered form in cosmetics, lubricants, and as an additive in paper manufacture. French chalk and potstone are varieties of talc. Soapstone has a greasy feel to it, and is used for carvings such as Inuit sculptures.

tectonics

The study of the movements of rocks on the earth's surface. On a small scale, tectonics involves the formation of folds and faults, but on a large scale plate tectonics deals with the movement of the earth's surface as a whole.

terminal moraine

A linear, slightly curved ridge of rocky debris deposited at the front end, or snout, of a glacier. It represents the furthest point of advance of a glacier, being formed when deposited material (till), which was pushed ahead of the snout as it advanced, became left behind as the glacier retreated. A terminal moraine may be hundreds of meters in height; for example, the Franz Joseph glacier in New Zealand has a terminal moraine that is over 400 m/1,320 ft high.

terrane

Also *terrain.* A tract of land with a distinct geological character. The term *exotic terrane* is commonly used to describe a rock mass that has a very different history from others near by. The exotic terranes of the Western Cordillera of North America (the broad mountainous region from the eastern edge of the Rocky Mountains to the Pacific Ocean) represent old island chains that have been brought to the North American continent by the movements of plate tectonics, and welded to its edge.

terrigenous

Derived from or pertaining to the land. River sediment composed of weathered rock material and deposited near the mouth of the river on the ocean's continental shelf (the shallow ledge extending out from the continent) is called *terrigenous sediment.*

Tertiary

The period of geological time 65 million to 1.64 million years ago, divided into five epochs: Paleocene, Eocene, Oligocene, Miocene, and Pliocene. During the Tertiary period, mammals took over all the ecological niches left vacant by the extinction of the dinosaurs, and became the prevalent land animals. The continents took on their present positions, and climatic and vegetation zones as we know them became established. Within the geological time column the Tertiary follows the Cretaceous period and is succeeded by the Quaternary period.

Tethys Sea

A sea that in the Mesozoic era separated Laurasia from Gondwanaland. The formation of the Alpine Mountains caused the sea to separate into the Mediterranean, the Black, the Caspian, and the Aral Seas.

thermosphere

The layer in the earth's atmosphere above the mesosphere and below the exosphere. Its lower level is about 80 km/50 mi above the ground, but its upper level is undefined. The ionosphere is located in the thermosphere. In the thermosphere the temperature rises with increasing height to several thousand degrees Celsius. However, because of the thinness of the air, very little heat is actually present.

thunderstorm

A severe storm of very heavy rain, thunder, and lightning. Thunderstorms are usually caused by the intense heating of the ground surface during summer. The warm air rises rapidly to form tall cumulonimbus clouds with a characteristic anvil-shaped top. Electrical charges accumulate in the clouds and are discharged to the ground as flashes of lightning. Air in the path of lightning becomes heated and expands rapidly, creating shock waves that are heard as a crash or rumble of thunder.

The rough distance between an observer and a lightning flash can be calculated by timing the number of seconds between the flash and the thunder. A gap of three seconds represents about a kilometer; five seconds represents about a mile.

tidal wave
The common name for a tsunami.

tide
The rhythmic rise and fall of the sea level in the earth's oceans and their inlets and estuaries due to the gravitational attraction of the moon and, to a lesser extent, the sun, affecting regions of the earth unequally as it rotates. Water on the side of the earth nearest the moon feels the moon's pull and accumulates directly below it, producing high tide.

High tide occurs at intervals of 12 hr 24 min 30 sec. The maximum high tides, or spring tides, occur at or near the new and full moon when the moon and sun are in line and exert the greatest combined gravitational pull. Lower high tides, or *neap tides,* occur when the moon is in its first or third quarter and the moon and sun are at right angles to each other.

till
Also *boulder clay.* A deposit of clay, mud, gravel, and boulders left by a glacier. It is unsorted, with all sizes of fragments mixed up together, and shows no stratification; that is, it does not form clear layers or beds.

time zone
A longitudinal strip of the earth's surface, stretching from Pole to Pole and sharing the same time of day or night. In a 24-hour period the earth makes one complete rotation on its axis; thus the direct rays of the sun pass through one degree of longitude every four minutes. To allow for time changes on an hourly basis, each time zone covers 15 degrees of longitude in width. In practice, however, zone boundary lines are adjusted to accommodate political units. (*See also* **international date line.**)

tin ore
A mineral from which tin is extracted, principally cassiterite, SnO_2. The world's chief producers are Malaysia, Thailand, and Bolivia.

titanium ore
Any mineral from which titanium is extracted, principally ilmenite ($FeTiO_3$) and rutile (TiO_2). Brazil, India, and Canada are major producers. Both these ore minerals are found either in rock formations or concentrated in heavy mineral sands.

topaz
A mineral, aluminum fluorosilicate, $Al_2(F_2SiO_4)$. It is usually yellow, but pink if it has been heated, and is used as a gemstone when transparent. It ranks 8 on the Mohs scale of hardness.

topography
The surface shape and composition of the landscape, comprising both natural and artificial features, and its study. Topographical features include the relief and

contours of the land; the distribution of mountains, valleys, and human settlements; and the patterns of rivers, roads, and railroads.

tor

An isolated mass of rock, often granite, left upstanding on a hilltop after the surrounding rock has been broken down. Weathering takes place along the joints in the rock, reducing the outcrop into a mass of rounded blocks.

tourmaline

A hard, brittle mineral, a complex silicate of various metals, but mainly sodium aluminum borosilicate, $(Na,Ca)(Mg,Fe^{2+},Fe^{3+},Al,Li)_3Al_6(BO_3)_3Si_6O_{18}(OH)_4$. Small tourmalines are found in granites and gneisses. The common varieties range from black (schorl) to pink, and the transparent gemstones may be colorless (achromatic), rose pink (rubellite), green (Brazilian emerald), blue (indicolite, verdelite, Brazilian sapphire), or brown (dravite).

trade wind

A prevailing wind that blows toward the equator from the northeast and southeast. Trade winds are caused by hot air rising at the equator and the consequent movement of air from north and south to take its place. The winds are deflected toward the west because of the earth's west-to-east rotation. The unpredictable calms known as the doldrums lie at their convergence. The trade wind belts move north and south about 5° with the seasons. The name is derived from the obsolete expression *blow trade* meaning to blow regularly, which indicates the trade winds' importance to navigation in the days of cargo-carrying sailing ships.

tremor

A minor earthquake.

Triassic

The period of geological time 245 to 208 million years ago, the first period of the Mesozoic era. The continents were fused together to form the world continent Pangaea. Triassic sediments contain remains of early dinosaurs and other reptiles now extinct. By late Triassic times, the first mammals had evolved. The climate was generally dry; desert sandstones are typical Triassic rocks.

troilite

The probable mineral of the earth's core, FeS, abundant in meteorites.

tropical cyclone

Another term for *hurricane.*

Tropics

The area between the Tropics of Cancer and Capricorn, defined by the parallels of latitude approximately 23° 30' N and S of the equator. They are the limits of the area of Earth's surface in which the sun can be directly overhead. The mean monthly temperature is over 20°C/68°F. Climates within the Tropics lie in parallel bands. Along the equator is the intertropical convergence zone, characterized by high temperatures and year-round heavy rainfall. Tropical rainforests are found here. Along the Tropics themselves lie the tropical high-pressure zones, characterized by descending dry air and desert conditions. Between these, the conditions vary seasonally between wet and dry, producing the tropical grasslands.

troposphere

The lower part of the earth's atmosphere extending about 10.5 km/6.5 mi from the earth's surface, in which temperature decreases with height to about -60°C/-76°F except in local layers of temperature inversion. The *tropopause* is the upper boundary of the troposphere, above which the temperature increases slowly with height within the atmosphere. All of the earth's weather takes place within the troposphere.

truncated spur

A blunt-ended ridge of rock jutting from the side of a glacial trough, or valley. As a glacier moves down a river valley it is unable to flow around the interlocking spurs that project from either side, and so it erodes straight through them, shearing away their tips and forming truncated spurs.

tsunami

Japanese for *harbor wave.* An ocean wave generated by vertical movements of the seafloor resulting from earthquakes or volcanic activity. Unlike waves generated by surface winds, the entire depth of water is involved in the wave motion. In the open ocean the tsunami takes the form of several successive waves, rarely in excess of 1 m/3 ft in height but traveling at speeds of 650–800 kph/400–500 mph. In the coastal shallows tsunamis slow down and build up, producing huge swells over 15 m/45 ft high in some cases and over 30 m/90 ft high in rare instances. The waves sweep inland causing great loss of life and property. On 26 May 1983 an earthquake in the Pacific Ocean caused tsunamis up to 14 m/42 ft high, which killed 104 people along the western coast of Japan near Minehama, Honshu. Before each wave there may be a sudden withdrawal of water from the beach. Used synonymously with tsunami, the popular term *tidal wave* is misleading: tsunamis are not caused by the gravitational forces that affect tides.

tufa

Also *travertine.* A soft, porous, limestone rock, white in color, deposited from solution from carbonate-saturated groundwater around hot springs and in caves. Undersea tufa columns, such as those in the Ikka Fjord in southwest Greenland, that form over alkaline springs, can reach 20 m/65 ft in height and grow at about 50 cm/20 in per year. They provide a habitat for a wide variety of marine life.

tundra

A region of high latitude almost devoid of trees, resulting from the presence of permafrost. The vegetation consists mostly of grasses, sedges, heather, mosses, and lichens. Tundra stretches in a continuous belt across northern North America and Eurasia. Tundra is also used to describe similar conditions at high altitudes. The term was originally applied to the topography of part of northern Russia, but is now used for all such regions.

tungsten ore

Either of the two main minerals, wolframite $(FeMn)WO_4$ and scheelite, $CaWO_4$, from which tungsten is extracted. Most of the world's tungsten reserves are in China, but the main suppliers are Bolivia, Australia, Canada, and the United States.

turbidity current

A gravity-driven current in air, water, or other fluid resulting from accumulation of suspended material, such as silt, mud, or volcanic ash, and imparting a density greater than the surrounding fluid. Marine turbidity currents originate from tectonic movement, storm waves, tsunamis (tidal waves), or earthquakes and move rapidly downward, like underwater avalanches, leaving distinctive deposits called *turbidites*. They are thought to be one of the mechanisms by which submarine canyons are formed.

turquoise

A mineral, hydrous basic copper aluminum phosphate, $CuAl_6(PO_4)_4(OH)_8 5H_2O$. Blue green, blue, or green, it is a gemstone. Turquoise is found in Australia, Egypt, Ethiopia, France, Germany, Iran, Turkestan, Mexico, and the southwestern United States. It was originally introduced into Europe through Turkey, from which its name is derived.

twilight

The period of faint light that precedes sunrise and follows sunset, caused by the reflection of light from the upper layers of the atmosphere. The limit of twilight is usually regarded as being when the sun is 18° below the horizon. The length of twilight depends on the latitude. At the Tropics, it only lasts a few minutes; near the Poles, it may last all night.

typhoon

A violent revolving storm, a hurricane in the western Pacific Ocean.

ultrabasic

An igneous rock with a lower silica content than basic rocks (less than 45 percent silica). Part of a system of classification based on the erroneous concept of silica acidity and basicity. Once used widely it has now been largely replaced by the term *ultramafic*.

unconformity

A surface of erosion or nondeposition eventually overlain by younger sedimentary rock strata and preserved in the geologic record. A surface where the beds above and below lie at different angles is called an *angular unconformity*. The boundary between older igneous or metamorphic rocks that are truncated by erosion and later covered by younger sedimentary rocks is called a *nonconformity.*

uniformitarianism

The principle that processes that can be seen to occur on the earth's surface today are the same as those that have occurred throughout geological time. For example, desert sandstones containing sand-dune structures must have been formed under conditions similar to those present in deserts today. The principle was formulated by James Hutton and expounded by Charles Lyell.

uraninite

Uranium oxide, UO_2, an ore mineral of uranium, also known as *pitchblende* when occurring in massive form. It is black or brownish black, very dense, and radioactive. It occurs in veins and as massive crusts, usually associated with granite rocks.

uranium ore

The material from which uranium is extracted, often a complex mixture of minerals. The main ore is uraninite (or pitchblende), UO_2, which is commonly found with sulfide minerals. The United States, Canada, and South Africa are the main producers in the West.

U-shaped valley

Also *glacial trough.* A valley formed by a glacier.

Valley of Ten Thousand Smokes

A valley in southwest Alaska, on the Alaska Peninsula, where in 1912 Mount Katmai erupted in one of the largest volcanic explosions ever known, although without loss of human life because the area was uninhabited. The valley was filled with ash to a depth of 200 m/660 ft. It was dedicated as the Katmai National Monument in 1918. Thousands of fissures on the valley floor continued to emit steam and gases for decades afterward.

varve

A pair of thin sedimentary beds, one coarse and one fine, representing a cycle of thaw followed by an interval of freezing, in lakes of glacial regions. Each couplet thus constitutes the sedimentary record of a year, and by counting varves in glacial lakes a record of absolute time elapsed can be determined. Summer and winter layers often are distinguished also by color, with lighter layers representing summer deposition, and darker layers being the result of dark clay settling from water while the lake was frozen.

veldt

A subtropical grassland in South Africa, equivalent to the Pampas of South America.

vermilion

A red form of mercuric sulfide, HgS; a scarlet that occurs naturally as the crystalline mineral cinnabar.

Vesuvius

Italian *Vesuvio.* An active volcano in Campania, Italy, 15 km/9 mi southeast of Naples, Italy, 1,277 m/4,190 ft in height. Vesuvius is a composite volcano at the convergent plate margin where the African plate is subducting beneath the Eurasian plate. Its lava is andesite in composition and consequently very viscous, giving rise to explosive eruptions. Vesuvius is comprised of two cones. Monte Somma, the remnant of a massive wall which once enclosed a huge cone in prehistoric times, is now a semicircular girdle of cliff to the north and east, separated from the main eruptive cone by the valley of Atrio di Cavallo. Layers of lava, scoriae, ashes, and pumice make up the mountain.

The surprising fertility of the volcano's slopes, especially for the cultivation of grapes and production of "Lacrimae Christi" wine, explains why the environs of Vesuvius remain densely populated in spite of the constant threat of eruption.

An eruption on 24 August A.D. 79 destroyed the cities of Pompeii, Herculaneum, and Stabiae and ended a dormant period so long that the volcano had been presumed extinct. During the eruptions of 472 and 1631 particles of dust are said to have landed in Constantinople (modern Istanbul). Other years of great activity were 1794, 1822, 1855, 1871, 1906, 1929, and 1944. There has been no eruption since 1944.

volcanic rock

Another name for extrusive rock, igneous rock formed on the earth's surface.

volcano

A crack in the earth's crust through which hot magma (molten rock) and gases well up. The magma is termed *lava* when it reaches the surface. A volcanic mountain, usually cone shaped with a crater on top, is formed around the opening, or vent, by the buildup of solidified lava and ashes (rock fragments). Most volcanoes arise on plate margins (*see* **plate tectonics**), where the movements of plates generate magma or allow it to rise from the mantle beneath. However, a number are found far from plate-margin activity, on "hot spots" where the earth's crust is thin.

There are two main types of volcano:

Composite volcanoes, such as Stromboli and Vesuvius in Italy, are found at destructive plate margins (areas where plates are being pushed together), usually in association with island arcs and coastal mountain chains. The magma is mostly derived from plate material and is rich in silica. This makes a very stiff lava such as andesite, which solidifies rapidly to form a high, steep-sided volcanic mountain. The magma often clogs the volcanic vent, causing violent eruptions as the blockage is blasted free, as in the eruption of Mount St. Helens in 1980. The crater may collapse to form a caldera.

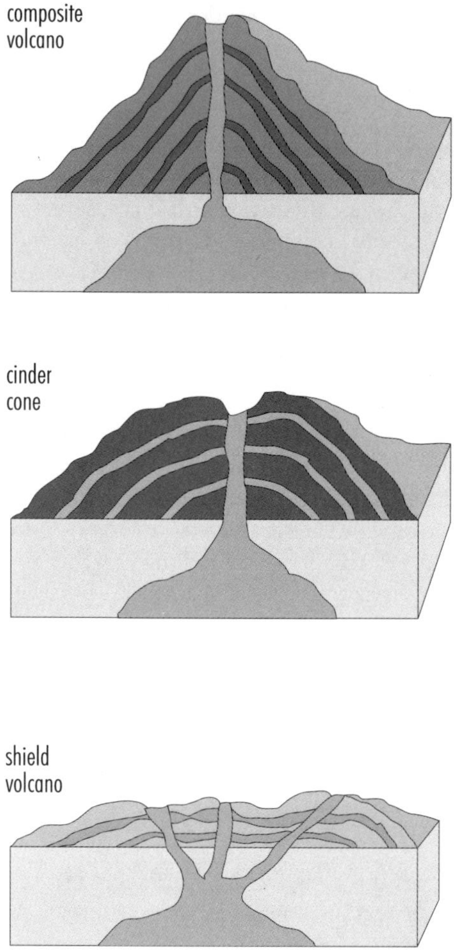

Shield volcanoes, such as Mauna Loa in Hawaii, are found over hot spots and also along the rift valleys and ocean ridges of constructive plate margins (areas where plates are moving apart). The magma is derived from the earth's mantle and is quite free flowing. The lava formed from this magma—usually basalt—flows for some distance over the surface before it sets and so forms broad low volcanoes. The lava of a shield volcano is not ejected violently but simply flows over the crater rim.

The type of volcanic activity is also governed by the age of the volcano. The first stages of an eruption are usually vigorous as the magma forces its way to the surface. As the pressure drops and the vents become established, the main phase of activity begins, composite volcanoes giving pyroclastic debris and shield volcanoes giving lava flows. When the pressure from below ceases, due to exhaustion of the magma chamber, activity wanes and is confined to the emission of gases and in time this also ceases. The volcano then enters a period of quiescence, after which activity may resume after a period of days, years, or even thousands of years. Only when the root zones of a volcano have been exposed by erosion can a volcano be said to be truly extinct.

Many volcanoes are submarine and occur along midoceanic ridges. The chief terrestrial volcanic regions are around the Pacific Rim (Cape Horn to Alaska); the central Andes of Chile (with the world's highest volcano, Guallatiri, 6,060 m/ 19,900 ft); North Island, New Zealand; Hawaii; Japan; and Antarctica. There are more than 1,300 potentially active volcanoes on Earth. Volcanism has helped shape other members of the solar system, including the moon, Mars, Venus, and Jupiter's moon Io.

There are several methods of monitoring volcanic activity. They include seismographic instruments on the ground, aircraft monitoring, and space monitoring using remote-sensing satellites.

V-shaped valley

A river valley with a V-shaped cross section. Such valleys are usually found near the source of a river, where the steeper gradient means that there is a great deal of corrasion (grinding away by rock particles) along the stream bed and erosion cuts downward more than it does sideways. However, a V-shaped valley may also be formed in the lower course of a river when its powers of downward erosion become renewed by a fall in sea level, a rise in land level, or the capture of another river (*see* **rejuvenation**).

vulcanology

The study of volcanoes and the geological phenomena that cause them.

wadi

In arid regions of the Middle East, a steep-sided valley containing an intermittent stream that flows in the wet season.

Waldsterben

German for *forest death.* A tree decline related to air pollution, common throughout the industrialized world. It appears to be caused by a mixture of pollutants; the precise chemical mix varies among locations, but it includes acid rain, ozone, sulfur dioxide, and nitrogen oxides.

Waldsterben was first noticed in the Black Forest of Germany during the late 1970s, and is spreading to many Third-World countries, such as China. Research has now shown Britain's trees to be among the most badly affected in Europe. Only 6 percent of the trees in Britain were undamaged by pollution in 1991, and about 57 percent had lost more than a quarter of their leaves. Acid rain is the main cause of this damage.

water cycle

Also *hydrological cycle.* The natural circulation of water through the biosphere. It is a complex system involving a number of physical and chemical processes (such as evaporation, precipitation, and infiltration) and stores (such as rivers, oceans, and soil).

Water is lost from the earth's surface to the atmosphere by evaporation caused by the sun's heat on the surface of lakes, rivers, and oceans, and through the transpiration of plants. This atmospheric water is carried by the air moving across the earth, and condenses as the air cools to form clouds, which in turn deposit moisture on the land and sea as precipitation. The water that collects on land flows to the ocean overland—as streams, rivers, and glaciers—or through the soil (infiltration) and rock (groundwater). The boundary that marks the upper limit of groundwater is called the *water table.* The oceans, which cover around 70 percent of the earth's surface, are the source of most of the moisture in the atmosphere.

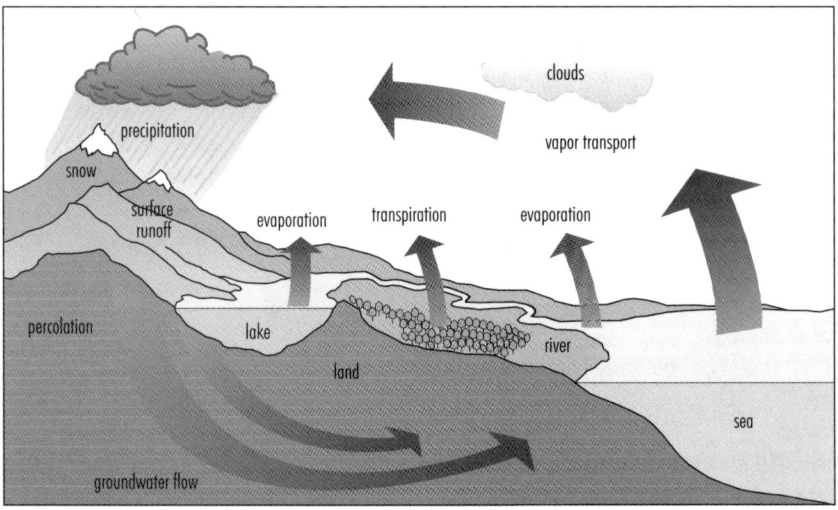

water table

The upper level of groundwater (water collected underground in porous rocks). Water that is above the water table will drain downward; a spring forms where the water table cuts the surface of the ground. The water table rises and falls in response to rainfall and the rate at which water is extracted, for example, for irrigation and industry. In many irrigated areas the water table is falling due to the extraction of water. Below northern China, for example, the water table is sinking at a rate of 1 m/3 ft a year. Regions with high water tables and dense industrialization have problems with pollution of the water table. New Jersey, Florida, and Louisiana have water tables contaminated by both industrial wastes and saline seepage from the ocean.

waterfall

A cascade of water in a river or stream. It occurs when a river flows over a bed of rock that resists erosion; weaker rocks downstream are worn away, creating a steep, vertical drop and a plunge pool into which the water falls. Over time, continuing erosion causes the waterfall to retreat upstream forming a deep valley, or gorge.

waterspout

A funnel-shaped column of water and cloud that is drawn from the surface of the sea or a lake by a tornado.

wave

In the oceans, a ridge or swell formed by wind or other causes. The power of a wave is determined by the strength of the wind and the distance of open water over which the wind blows (the fetch). Waves are the main agents of coastal erosion and deposition; sweeping away or building up beaches, creating spits and berms, and wearing down cliffs by their hydraulic action and by the corrosion of the sand and gravel that they carry. A tsunami (misleadingly called a *tidal wave*) is formed after a submarine earthquake.

As a wave approaches the shore it is forced to break. Friction with the sea bed causes the wavelenth to decrease, while the shallow depth causes the wave height to increase. The wave eventually becomes unstable and breaks. When it breaks on a beach, water and sediment are carried up the beach as swash; the water then drains back as backwash.

A *constructive wave* causes a net deposition of material on the shore because its swash is stronger than its backwash. Such waves tend be low and have crests that spill over gradually as they break. The backwash of a *destructive wave* is stronger than its swash, and therefore causes a net removal of material from the shore. Destructive waves are usually tall and have peaked crests that plunge downward as they break, trapping air as they do so.

If waves strike a beach at an angle, the beach material will be gradually moved along the shore (longshore drift), causing a deposition of material in some areas and erosion in others.

Atmospheric instability caused by global warming, possibly due to the greenhouse effect, appears to be increasing the severity of Atlantic storms and affecting the heights of the ocean waves. Waves in the South Atlantic are shrinking—they are on average half a meter smaller than in the mid-1980s—and those in the northeast Atlantic have doubled in size over the last 40 years. As the height of waves affects the supply of marine food, this could affect fish stocks, and there are also implications for shipping and oil and gas rigs in the North Atlantic, which will need to be strengthened if they are to avoid damage.

Freak or *episodic* waves form under particular weather conditions at certain times of the year, traveling long distances in the Atlantic, Indian, and Pacific Oceans. They are considered responsible for the sudden disappearance, without distress calls, of many ships. Freak waves become extremely dangerous when they reach the shallow waters of the continental shelves at 100 fathoms (180 m/600 ft), especially when they meet currents: for example, the Agulhas Current to the east of South Africa, and the Gulf Stream in the North Atlantic. A wave height of 34 m/112 ft has been recorded.

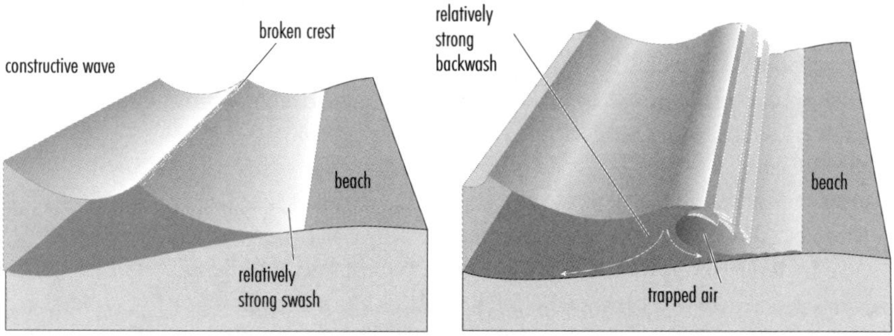

wave refraction

The distortion of waves as they reach the coast, due to variations in the depth of the water and shape of the coastline. It is particularly evident where there are headlands and bays. The bending of a wave crest as it approaches a headland concentrates the energy of the wave in the direction of that headland, and increases its power of erosion. By contrast, the bending that a wave crest experiences when it moves into a bay causes its energy to be dissipated away from the direction of the shore. As a result the wave loses its erosive power and becomes more likely to deposit sediment on the shore.

wave-cut platform

A gently sloping rock surface found at the foot of a coastal cliff. Covered by water at high tide but exposed at low tide, it represents the last remnant of an eroded headland (*see* **coastal erosion**).

weather

The day-to-day variation of atmospheric and climatic conditions at any one place over a short period of time. Such conditions include humidity, precipitation, temperature, cloud cover, visibility, and wind, together with extreme phenomena such as storms and blizzards. Weather differs from climate in that the latter is a composite of the average weather conditions of a locality or region over a long period of time (at least 30 years). Meteorology is the study of short-term weather patterns and data within a circumscribed area; climatology is the study of weather over longer time scales on a zonal or global basis.

Weather forecasts. Forecasts are based on current meteorological data, and predict likely weather for a particular area; they may be short range (covering a period of one or two days), medium range (five to seven days), or long range (a month or so). Weather observations are made on an hourly basis at meteorological recording stations. There are more than 3,500 of these around the world. More than 140 nations participate in the exchange of weather data through the World Weather Watch program, which is sponsored by the World Meteorological Organization (WMO), and information is distributed among the member nations by means of a worldwide communications network. Incoming data is collated at weather centers in individual countries and plotted on weather maps, or charts. The weather map uses internationally standardized symbols to indicate barometric pressure, cloud cover, wind speed and direction, precipitation, and other details reported by each recording station at a specific time. Points of equal atmospheric pressure are joined by lines called *isobars* and from these the position and movement of weather fronts and centers of high and low pressure can be extrapolated. The charts are normally compiled on a three-hourly or six-hourly basis; the main synoptic hours are midnight, 0600, 1200 and 1800. Predictions for future weather are drawn up on the basis of comparisons between current charts and previous charts. Additional data received from weather balloons and satellites help to complete and corroborate the picture obtained from the weather map.

In the early days of weather forecasting, when few recording stations were available, the features that stood out most clearly on weather maps were the patterns made by isobars, forming regions of low pressure (depressions or cyclones), regions of high pressure (anticyclones), and the connecting patterns: ridges, troughs, cols, and secondary depressions.

One of the founders of meteorology as a science was the Dutch mathematician, physicist, and meteorologist Christoph Hendrik Diederik Buys Ballot (1817–1890), who was a professor at the University of Utrecht. He advocated international cooperation in the field of meteorology and in 1854 founded the Royal Netherlands Meteorological Institute. The invention of the telegraph made the collation of simultaneous weather observations possible, and enabled Buys Ballot to set up a wide network for the exchange of weather data. Nowadays, he is best known for his observation of 1857, known as *Buys Ballot's law,* that if one stands with one's back to the wind in the Northern Hemisphere, atmospheric pressure will be lower on the left than on the right, and the converse in the Southern Hemisphere. In fact, unknown to Buys Ballot, a theory to this effect had already been put forward by an American meteorologist, William Ferrel (1817–1891), who expanded on this with a theory on the deflection of air currents on the rotating Earth.

Gradient and geostrophic winds. Ferrel's law was confirmed by the formula that gave a theoretical speed and direction to the wind in the free atmosphere, known as the *gradient wind.* In the Northern Hemisphere, with low pressure on the left, this wind blows along the isobars with a speed proportional to the pressure gradient between the isobars (the closer together the isobars, the stronger the wind). It is increased by a smaller quantity for anticyclonic curvature, and decreased similarly for cyclonic curvature of the isobars. The effect of the curved isobars is normally small enough to neglect; this wind is known theoretically as the *geostrophic wind.*

However, both the gradient and geostrophic winds neglect to take into account the effect of change of pressure with time and vertical motion, as well as frictional forces. These values are all generally small and the wind speed at about 600 m/1,970 ft above the surface may be read off the synoptic chart by placing a scale appropriate to the latitude across the isobars. Surface wind is affected considerably by friction and the configuration of the land, especially in hilly regions where it tends to blow along the valleys; the speed is usually less than the gradient wind, and the direction at an angle (often 20–30°) across the isobars from high pressure to low. The effects neglected by the gradient wind equation are important, for it is these that lead to inflow of surface air to regions of low or falling pressure and, conversely, of outflow from regions of high and rising pressure. If the surface air is flowing into a region, then the air must escape by rising and spreading out aloft; this leads to adiabatic cooling and instability, clouds, and unsettled weather. Conversely, diverging air leads to subsidence, adiabatic warming, stability, clear skies, and settled weather. These theories are, in general, confirmed by the observation of unsettled weather in depressions and settled weather in anticyclones.

Orographic rain. When air blows over hills and mountains it is forced to rise and, if it contains enough moisture, cloud is formed; if the process continues long enough, or the air is very moist or unstable, rain falls. This type of rain is called *orographic* rain, and tends to be prevalent on coasts exposed to frequent winds from the sea. Even if the humidity of the air is not sufficient to produce rain, drizzle often occurs, and even very low hills may be shrouded in clouds; these are the conditions that produce a mist in the Western Isles of Scotland or the *crachin* of South China. Areas on the leeside of the hills are likely to experience finer and drier weather as the air will have lost most of its moisture before reaching them and many of the clouds will have dispersed.

Weather fronts. During World War I, Norwegian meteorologists Vilhelm Bjerknes (1862–1951) and his son Jacob (1897–1975), using a very close network of reporting stations, noticed that the change from one relatively homogenous type of air mass to another was often very rapid with clear demarcations, which they termed fronts. They found that these frontal zones are regions of widespread cloud development and precipitation, and that the air masses exhibit specific changes in weather characteristics as they move from colder to warmer regions and vice versa. As an air mass moves to a warmer region, the lower atmosphere becomes warm and the air becomes unstable, leading to convective cumulus cloud and showers; on the other hand, when an air mass moves over colder regions the lower air becomes colder, resulting in condensation and the development of fog or a very low layered cloud known as *stratus.* From these observations Jacob Bjerknes produced the theory that the atmosphere consists of warm (tropical) and cold (polar) air masses separated by distinct boundaries (fronts) and that it is wave disturbances in these fronts that give rise to cyclones or depressions. This theory greatly aided the accuracy of weather forecasting. In 1939 Bjerknes emigrated to the United States where he contributed significantly to the war effort by training meteorologists for the U.S. aviation services.

Modern forecasting. Meteorologists, aided by communications and computer technology, are increasingly able to refine the accuracy of forecasts, but however sophisticated the techniques become, there will always be an element of the unknown in any forecast that, as shown by the chaos theory (the branch of mathematics that attempts to describe irregular, unpredictable systems), is unlikely to be eliminated. This is often referred to as the *butterfly effect;* the flapping of a butterfly's wings, or a small gust of wind, on one side of the earth will have an ongoing effect that will eventually be felt on the opposite side of the globe.

weather vane

An instrument that shows the direction of wind. Wind direction is always given as the direction from which the wind has come; for example, a northerly wind comes from the north.

westerlies

Prevailing winds from the west that occur in both hemispheres between latitudes of about 35° and 60°. Unlike the trade winds, they are very variable and produce stormy weather. The westerlies blow mainly from the southwest in the Northern Hemisphere and the northwest in the Southern Hemisphere, bringing moist weather to the western coasts of the landmasses in these latitudes.

wetland

A permanently wet land area or habitat. Wetlands include areas of marsh, fen, bog, flood plain, and shallow coastal areas. Wetlands are extremely fertile. They provide warm sheltered waters for fisheries, lush vegetation for grazing livestock, and an abundance of wildlife. Estuaries and seaweed beds are more than 16 times as productive as the open ocean.

The term is often more specifically applied to a naturally flooding area that is managed for agriculture or wildlife. A water meadow, where a river is expected to flood grazing land at least once a year, thereby replenishing the soil, is a traditional example.

The largest area of tidal wetlands in Japan, a 3,550-hectare area in Isahaya Bay, near Nagasaki, is threatened by a 7 km barrier erected across the bay in 1997 as part of a flood-control project. The wetlands, cut off from the tidal waters, are drying out rapidly and the ecosystem is likely to die if the barrier is not removed immediately. The mudflats are home to 282 species of birds, crustaceans, and fish, including 21 endangered species.

wetted perimeter

The length of that part of a river's cross section that is in contact with the water. The wetted perimeter is used to calculate a river's hydraulic radius, a measure of its channel efficiency.

whirlwind

A rapidly rotating column of air, often synonymous with a tornado. On a smaller scale it produces the dust devils seen in deserts.

whiteout

A "fog" of grains of dry snow caused by strong winds in temperatures of between -18°C/0°F and -1°C/30°F. The uniform whiteness of the ground and air causes disorientation in humans.

willy-willy

An Australian Aboriginal term for a cyclonic whirlwind.

wind

The lateral movement of the earth's atmosphere from high-pressure areas (anticyclones) to low-pressure areas (depressions). Its speed is measured using an anemometer or by studying its effects on, for example, trees by using the Beaufort scale. Although modified by features such as land and water, there is a basic worldwide system of trade winds, westerlies, and polar easterlies.

A belt of low pressure (the doldrums) lies along the equator. The trade winds blow toward this from the horse latitudes (areas of high pressure at about 30° N and 30° S of the equator), blowing from the northeast in the Northern Hemisphere, and from the southeast in the Southern Hemisphere. The westerlies (also from the horse latitudes) blow north of the equator from the southwest and south of the equator from the northwest.

Cold winds blow outward from high-pressure areas at the Poles. More local effects result from landmasses heating and cooling faster than the adjacent sea, producing onshore winds in the daytime and offshore winds at night.

The monsoon is a seasonal wind of southern Asia, blowing from the southwest in summer and bringing the rain on which crops depend. It blows from the northeast in winter.

Famous or notorious warm winds include the chinook of the eastern Rocky Mountains, North America; the föhn of Europe's Alpine valleys; the sirocco of Italy or khamsin in Egypt, and sharav in Israel, which are spring winds that bring warm air from the Sahara and Arabian Deserts across the Mediterranean; and the Santa Ana, a periodic warm wind from the inland deserts that strikes the California coast. The dry northerly bise (Switzerland) and the mistral, which strikes the Mediterranean area of France, are unpleasantly cold winds.

wolframite

Iron manganese tungstate, $(Fe,Mn)WO_4$, an ore mineral of tungsten. It is dark gray with a submetallic surface luster, and often occurs in hydrothermal veins in association with ores of tin.

woodland

An area in which trees grow more or less thickly; generally smaller than a forest. Temperate climates, with four distinct seasons a year, tend to support a mixed woodland habitat, with some conifers but mostly broad-leaved and deciduous trees, shedding their leaves in autumn and regrowing them in spring. In the Mediterranean region and parts of the Southern Hemisphere, the trees are mostly evergreen. Temperate woodlands grow in the zone between the cold coniferous forests and the tropical forests of the hotter climates near the equator. They develop in areas where the closeness of the sea keeps the climate mild and moist.

Old woodland can rival tropical rainforest in the number of species it supports, but most of the species are hidden in the soil. A study in Oregon in 1991 found that the soil in a single woodland location contained 8,000 arthropod species (such as insects, mites, centipedes, and millipedes), compared with only 143 species of reptile, bird, and mammal in the woodland above.

In England in 1900 about 2.5 percent of the land was woodland, compared to about 3.4 percent in the eleventh century. An estimated 33 percent of ancient woodland has been destroyed since 1945.

The trees determine the character of the wood. Sometimes a single species dominates, as in a pine or beech wood, but there is often a mixture of two or more codominants, as in a mixed oak and ash wood. Beneath the tree canopy there is frequently a layer of shrubs and beneath these are the herbs. Woodland herbs grow in shady conditions and are adapted in various ways to make the best possible use of the available sunlight. The woodland floor provides moist conditions in which mosses and liverworts thrive and many fungi grow in the soil or on rotting wood. The trees themselves provide habitats for another group including climbing plants, mosses, liverworts, lichens, fungi, and microscopic algae.

yardang

A ridge formed by wind erosion from a dried-up riverbed or similar feature, as in Chad, China, Peru, and North America. On the planet Mars yardangs occur on a massive scale.

Yixian formation

In paleontology, a Chinese geological formation in the rural province of Liaoning that is yielding a wealth of extraordinarily well-preserved fossils. The fossils date from 150 million to 120 million years ago (late Jurassic or early Cretaceous) and include those of hundreds of early birds, such as *Confuciusornis,* with feathers, lizards with full skin, and mammals with hair.

zeolite

Any of the hydrous aluminum silicates, also containing sodium, calcium, barium, strontium, or potassium, chiefly found in igneous rocks and characterized by a ready loss or gain of water. Zeolites are used as "molecular sieves" to separate mixtures because they are capable of selective absorption. They have a high ion-exchange capacity and can be used to make gasoline, benzene, and toluene from low-grade raw materials, such as coal and methanol. Permutit is a synthetic zeolite used to soften hard water.

zinc ore

A mineral from which zinc is extracted, principally sphalerite (Zn,Fe)S, but also zincite, ZnO_2, and smithsonite, $ZnCO_3$, all of which occur in mineralized veins.

Ores of lead and zinc often occur together, and are common worldwide; Canada, the United States, and Australia are major producers.

zircon

Zirconium silicate, $ZrSiO_4$, a mineral that occurs in small quantities in a wide range of igneous, sedimentary, and metamorphic rocks. It is very durable and is resistant to erosion and weathering. It is usually colored brown, but can be other colors, and when transparent may be used as a gemstone. Zircons contain abundant radioactive isotopes of uranium and so are useful for uranium-lead dating to determine the ages of rocks.

Major Twentieth-Century Earthquakes

Date	Location	Magnitude (Richter scale)	Estimated deaths
April 18–19, 1906	San Francisco (CA), United States	7.7–7.9	503
August 16, 1906	Valparaiso, Chile	8.6	20,000
December 28, 1908	Messina, Italy	7.5	83,000
January 13, 1915	Avezzano, Italy	7.5	29,980
December 16, 1920	Gansu Province, China	8.6	200,000
September 1, 1923	Yokohama, Japan	8.3	143,000
May 22, 1927	Nan-Shan, China	8.3	200,000
December 26, 1932	Gansu, China	7.6	70,000
May 31, 1935	Quetta, India	7.5	60,000
January 24, 1939	Chillan, Chile	7.8	30,000
December 26, 1939	Erzincan, Turkey	7.9	23,000
December 21, 1946	Honshu, Japan	8.4	2,000
June 28, 1948	Fukui, Japan	7.3	5,130
October 6, 1948	Iran/USSR	7.3	100,000
August 5, 1949	Pelileo, Ecuador	6.8	6,000
August 15, 1950	Assam, India	8.7	1,530
March 18, 1953	northwestern Turkey	7.2	1,200
June 10–17, 1956	northern Afghanistan	7.7	2,000
July 2, 1957	northern Iran	7.4	2,500
December 13, 1957	western Iran	7.1	2,000
February 29, 1960	Agadir, Morocco	5.7	12,000

Date	Location	Magnitude (Richter scale)	Estimated deaths
September 1, 1962	northwestern Iran	7.1	12,000
July 26, 1963	Skopje, Yugoslavia	6.0	1,100
March 27, 1964	Anchorage (AL), United States	9.2	131
August 19, 1966	eastern Turkey	6.9	2,520
August 31, 1968	northeastern Iran	7.4	11,600
January 5, 1970	Yunan Province, China	7.7	10,000
May 31, 1970	Chimbote, Peru	7.8	67,000
April 10, 1972	southern Iran	7.1	5,000
December 23, 1972	Managua, Nicaragua	6.2	5,000
December 28, 1974	Kashmir, Pakistan	6.3	5,200
February 4, 1976	Guatemala City, Guatemala	7.5	22,778
May 6, 1976	Friuli, Italy	6.5	939
July 28, 1976	Tangshan, China	8.0	255,000[1]
March 4, 1977	Romania	7.5	1,541
September 16, 1978	northeastern Iran	7.7	25,000
December 12, 1979	Colombia/Ecuador	7.9	800
October 10, 1980	northern Algeria	7.7	3,000
November 23, 1980	southern Italy	7.2	4,800
December 13, 1982	northern Yemen	6.0	1,600
October 30, 1983	eastern Turkey	6.9	1,300
September 21, 1985	Mexico City, Mexico	8.1	5,000[2]
August 20, 1988	Nepal/India	6.9	1,000
November 6, 1988	southwestern China	7.6	1,000
December 7, 1988	Armenia, USSR	6.8	25,000
October 17, 1989	San Francisco (CA), United States	7.1	62
June 20–21, 1990	northwestern Iran	7.7	37,000
July 16, 1990	Luzon, Philippines	7.7	1,660
February 1, 1991	Afghanistan/Pakistan	6.8	1,000

[1]Early estimates put the death toll as high as 750,000; the figure shown is the official one.
[2]Some estimates put the death toll as high as 20,000.

Date	Location	Magnitude (Richter scale)	Estimated deaths
April 1991	northern Georgia	7.2	>100
October 20, 1991	Uttar Pradesh, India	6.1	1,500
March 15, 1992	Erzincan, Turkey	6.7	2,000
December 12, 1992	Flores Island, Indonesia	7.5	2,500
July 12, 1993	western coast of Hokkaido, Japan	7.8	200
September 29, 1993	Maharashtra, India	6.3	9,800
October 13–16, 1993	Papua New Guinea	6.8	>60
June 6, 1994	Cauca, Colombia	6.8	1,000
August 19, 1994	northern Algeria	5.6	200
January 16, 1995	Kobe, Japan	7.2	5,500
June 14, 1995	Sakhalin Island, Russia	7.6	2,000
October 2, 1995	southwestern Turkey	6.0	84
October 7, 1995	Sumatra, Indonesia	7.0	>70
October 9, 1995	Mexico	7.6	>66
February 3, 1996	Yunnan Province, China	7.0	>250
February 17, 1996	Irian Jaya, Indonesia	7.5	108
March 28, 1996	Ecuador	5.7	21
4, February 28, 1997	Ardabil, Iran	N/A	>1,000
February 28, 1997	Baluchistan Province, Pakistan	7.3	>100
May 10, 1997	northeastern Iran (Khorasah Province)	7.1	>1,600
May 22, 1997	India	6.0	>40
September 26, 1997	central Italy	5.8	>11
September 28, 1997	Sulawesi, Indonesia	6.0	>20
October 14, 1997	north of Santiago, Chili	6.8	>10
November 21, 1997	Chittagong, Bangladesh	6.0	17
January 11, 1998	northeastern China	6.2	>47
February 4, 1998	Takhar province, Afghanistan	6.1	>3,800

(N/A = not available.)

Major Twentieth-Century Hurricanes, Typhoons, Cyclones, and Other Storms

Date	Event/name	Location	Estimated deaths
August 27–September 15, 1900	hurricane	Galveston (TX), United States	6,000
September 18, 1906	typhoon	Hong Kong	10,000
September 19–24, 1906	hurricane	Louisiana and Mississippi, United States	350
August 5–23, 1915	hurricane	eastern Texas and Louisiana, United States	275
September 2–15, 1919	hurricane	Louisiana, Florida, and Texas, United States	775
March 18, 1925	tornadoes	midwestern United States	800
September 11–22, 1926	hurricane	Florida and Alabama, United States	243
October 20, 1926	hurricane	Cuba	600
September 6–20, 1928	hurricane	southern Florida, United States	1,836
September 3, 1930	hurricane	Dominican Republic	2,000
March 21, 1932	tornadoes	southern United States	268
August 29–September 10, 1935	hurricane	southern Florida, United States	408
April 5–6, 1936	tornadoes	southern United States	498
September 10–22, 1938	hurricane	east coast, United States	600
November 11–12, 1940	blizzard	northeastern and midwestern United States	144
October 16, 1942	cyclone	India	40,000
September 9–16, 1944	hurricane	east coast, United States	390

Date	Event/name	Location	Estimated deaths
September 4–21, 1947	hurricane	Florida and mid-Gulf Coast, United States	51
December 26, 1947	blizzard	New York and North Atlantic states, United States	55
March 21–22, 1952	tornadoes	southern United States	343
October 22, 1952	typhoon	Philippines	440
May 11, 1953	tornado	Waco (TX), United States	114
June 8, 1953	tornado	Flint (MI), United States	116
August 25–31, 1954	hurricane, *Carol*	northeastern United States	68
October 5–18, 1954	hurricane, *Hazel*	eastern United States/Haiti	347
August 7–12, 1955	hurricane, *Diane*	eastern United States	400
August 12–13, 1955	hurricane, *Connie*	Carolinas, Virginia, and Maryland, United States	43
September 19, 1955	hurricane, *Hilda*	Mexico	200
September 22–28, 1955	hurricane, *Janet*	Caribbean	500
February 1–29, 1956	blizzard	Europe	1,000
June 25–30, 1957	hurricane, *Audrey*	Texas to Alabama, United States	390
February 5–16, 1958	blizzard	northeastern United States	171
September 17–19, 1959	typhoon, *Sarah*	Japan/South Korea	2,000
September 26–27, 1959	typhoon, *Vera*	Japan	4,466
September 4–12, 1960	hurricane, *Donna*	Caribbean/eastern United States	148
October 10, 1960	cyclone	eastern Pakistan	6,000
September 11–14, 1961	hurricane, *Carla*	Texas, United States	46
October 31, 1961	hurricane, *Hattie*	British Honduras	400
May 28–29, 1963	cyclone	Bangladesh	22,000
October 4–8, 1963	hurricane, *Flora*	Caribbean	6,000
June 30, 1964	typhoon, *Winnie*	northern Philippines	107
September 5, 1964	typhoon, *Ruby*	Hong Kong/China	735
April 11, 1965	tornadoes	midwestern United States	256
May 11–12, 1965	cyclone	Bangladesh	17,000
June 1–2, 1965	cyclone	Bangladesh	30,000

Date	Event/name	Location	Estimated deaths
December 15, 1965	cyclone	Karachi, Pakistan	10,000
June 4–10, 1966	hurricane, *Alma*	Honduras/eastern United States	51
September 24–30, 1966	hurricane, *Inez*	Caribbean/Mexico/southeastern United States	293
July 9, 1967	typhoon, *Billie*	southwestern Japan	347
September 5–23, 1967	hurricane, *Beulah*	Caribbean/Mexico/Texas, United States	54
December 12–20, 1967	blizzard	southwestern United States	51
November 18–20, 1968	typhoon, *Nina*	Philippines	63
August 17–18, 1969	hurricane, *Camille*	Louisiana and Mississippi, United States	256
July 30–August 5, 1970	hurricane, *Celia*	Cuba/Florida and Texas, United States	31
August 20–21, 1970	hurricane, *Dorothy*	Martinique	42
September 15, 1970	typhoon, *Georgia*	Philippines	300
October 14, 1970	typhoon, *Sening*	Philippines	583
October 15, 1970	typhoon, *Titang*	Philippines	526
November 12–13, 1970	cyclone	Bangladesh	>300,000
August 1, 1971	typhoon, *Rose*	Hong Kong	130
September 29, 1971	cyclone	Orissa State, India	10,000–25,000
June 19–29, 1972	hurricane, *Agnes*	eastern United States	118
December 3, 1972	typhoon, *Theresa*	Philippines	169
April 3–4, 1974	tornadoes	eastern, southern, and midwestern United States	315
June 11, 1974	storm, *Dinah*	Luzon, Philippines	71
July 11, 1974	typhoon, *Gilda*	Japan/South Korea	108
September 19–20, 1974	hurricane, *Fifi*	Honduras	2,000
December 25, 1974	cyclone	Darwin, Australia	50
September 13–27, 1975	hurricane, *Eloise* United States	Caribbean/northeastern	71
May 20, 1976	typhoon, *Olga*	Philippines	215
July 31, 1977	typhoons, *Thelma*, *Vera*	Taiwan	39

Date	Event/name	Location	Estimated deaths
November 19, 1977	cyclone	Andhra Pradesh, India	20,000
October 27, 1978	typhoon, *Rita*	Philippines	>400
August 30–September 7, 1979	hurricane, *David*	Caribbean/eastern United States	1,100
August 4–11, 1980	hurricane, *Allen*	Caribbean/Texas, United States	272
November 25, 1981	typhoon, *Irma*	Luzon, Philippines	176
August 18, 1983	hurricane, *Alicia*	southern Texas, United States	21
September 2, 1984	typhoon, *Ike*	southern Philippines	1,363
May 25, 1985	cyclone	Bangladesh	10,000
October 26–November 6, 1985	hurricane, *Juan*	southeastern United States	63
November 25, 1987	typhoon, *Nina*	Philippines	650
September 10–17, 1988	hurricane, *Gilbert*	Gulf of Mexico/Caribbean	260
September 10–22, 1989	hurricane, *Hugo*	Caribbean/United States	86
March 12–18, 1990	storms	Bangladesh	242
May 6–11, 1990	cyclone	India	514
June 1990	typhoons	Philippines	156
July–August 1990	typhoons	China	1,802
November 1990	typhoons/storm	Philippines	1,312
April 30, 1991	cyclone	Bangladesh	138,866
May 12, 1991	storm	Philippines	3,956
June 10, 1991	cyclone	Bangladesh	125,720
February 15, 1992	cyclone	Vietnam	251
August 23–26, 1992	hurricane, *Andrew*	southern Florida and Louisiana, United States	>60
March 13–14, 1993	blizzard	eastern United States	270
May 2, 1994	cyclone	southeastern Bangladesh	165
August 22, 1994	typhoon, *Fred*	Zhejiang Province, China	>710
October 31–November 3, 1994	cyclone	India	260
November 8–18, 1994	storm	Caribbean/Florida, United States	830
January 23, 1995	snowstorms	Kashmir, India	>200

Date	Event/name	Location	Estimated deaths
September 5–7, 1995	hurricane, *Luis*	Caribbean	14
2– November 3, 1995	typhoon, *Angela*	Philippines	722
May 13, 1996	tornado	Bangladesh	>600
July 27–28, 1996	hurricane, *Cèsar*	Panama/El Salvador/Costa Rica	50
July 31–August 1, 1996	typhoon, *Herb*	Taiwan	400
September 5–6, 1996	hurricane, *Fran*	North Carolina and Virginia, United States	36
November 4, 1996	cyclone	Andhra Pradesh, India	>1,000
January 4–5, 1997	storms	Brazil	68
May 19, 1997	cyclone	Bangladesh	112
September 27, 1997	cyclone	Bangladesh	>47
October 9, 1997	hurricane, *Pauline*/floods	Pacific coast of Mexico	128
November 2–3, 1997	typhoon, *Linda*	southern Vietnam/Thailand	>358
February 23, 1998	tornadoes	central Florida, United States	38
April 8, 1998	tornadoes	Alabama, Georgia, Mississippi, United States	>41

Richter Scale

The Richter scale is based on measurement of seismic waves, used to determine the magnitude of an earthquake at its epicenter. The magnitude of an earthquake differs from its intensity, measured by the Mercalli scale, which is subjective and varies from place to place for the same earthquake. The Richter scale was named for U.S. seismologist Charles Richter (1900–1985).

Magnitude	Relative amount of energy released	Examples	Year
1	1		
2	31		
3	960		
4	30,000	Carlisle, England (4.7)	1979
5	920,000	Wrexham, Wales (5.1)	1990
6	29,000,000	San Fernando (CA) (6.5)	1971
		northern Armenia (6.8)	1988
7	890,000,000	Loma Prieta (CA) (7.1)	1989
		Kobe, Japan (7.2)	1995
		Rasht, Iran (7.7)	1990
		San Francisco (CA) (7.7–7.9)[1]	1906
8	28,000,000,000	Tangshan, China (8.0)	1976
		Gansu, China (8.6)	1920
		Lisbon, Portugal (8.7)	1755
9	850,000,000,000	Prince William Sound (AK) (9.2)	1964

[1]Richter's original estimate of a magnitude of 8.3 has been revised by two recent studies carried out by the California Institute of Technology and the U.S. Geological Survey.

Major Twentieth-Century Volcanic Eruptions

Volcano	Location	Year	Estimated deaths
Santa María	Guatemala	1902	1,000
Pelée	Martinique	1902	28,000
Taal	Philippines	1911	1,400
Kelut	Java, Indonesia	1919	5,500
Vulcan, Rabaul	Papua New Guinea	1937	500
Lamington	Papua New Guinea	1951	3,000
St. Helens	United States	1980	57
El Chichon	Mexico	1982	1,880
Nevado del Ruiz	Colombia	1985	23,000
Lake Nyos	Cameroon	1986	1,700
Pinatubo	Luzon, Philippines	1991	639
Unzen	Japan	1991	39
Mayon	Philippines	1993	70
Grimsvötn[1]	Iceland	1996	0
Soufriere	Montserrat	1997	23
Merapi	Java, Indonesia	1998	38

[1]The eruption caused severe flooding and melted enough ice to create a huge subglacial lake.

Major Volcanoes Active in the Twentieth Century

Volcano	Height m/ft	Location	Date of last activity
Africa			
Cameroon	4,096/13,353	isolated mountain, Cameroon	1986
Nyiragongo	3,470/11,385	Virungu, Democratic Republic of Congo	1994
Nyamuragira	3,056/10,028	Democratic Republic of Congo	1998
Ol Doinyo Lengai	2,886/9,469	Tanzania	1993
Lake Nyos	918/3,011	Cameroon	1986
Erta-Ale	503/1,650	Ethiopia	1995
Antarctica			
Erebus	4,023/13,200	Ross Island, McMurdo Sound	1995
Deception Island	576/1,890	South Shetland Island	1970
Asia			
Kerinci	3,800/12,467	Sumatra, Indonesia	1987
Rindjani	3,726/12,224	Lombok, Indonesia	1966
Semeru	3,676/12,060	Java, Indonesia	1995
Slamet	3,428/11,247	Java, Indonesia	1989
Raung	3,322/10,932	Java, Indonesia	1993
Agung	3,142/10,308	Bali, Indonesia	1964
On-Taka	3,063/10,049	Honshu, Japan	1991
Merapi	2,911/9,551	Java, Indonesia	1998

As of January 15, 1999.

Volcano	Height m/ft	Location	Date of last activity
Marapi	2,891/9,485	Sumatra, Indonesia	1993
Asama	2,530/8,300	Honshu, Japan	1990
Nigata Yake-yama	2,475/8,111	Honshu, Japan	1989
Mayon	2,462/8,084	Luzon, Philippines	1993
Canlaon	2,459/8,070	Negros, Philippines	1993
Chokai	2,225/7,300	Honshu, Japan	1974
Galunggung	2,168/7,113	Java, Indonesia	1984
Azuma	2,042/6,700	Honshu, Japan	1977
Sangeang Api	1,935/6,351	Lesser Sunda Island, Indonesia	1988
Pinatubo	1,759/5,770	Luzon, Philippines	1995
Kelut	1,730/5,679	Java, Indonesia	1990
Unzen	1,360/4,462	Japan	1996
Krakatoa	818/2,685	Sumatra, Indonesia	1996
Taal	300/984	Philippines	1977
Atlantic Ocean			
Pico de Teide	3,716/12,192	Tenerife, Canary Islands, Spain	1909
Fogo	2,835/9,300	Cape Verde Islands	1995
Beerenberg	2,277/7,470	Jan Mayen Island, Norway	1985
Hekla	1,491/4,920	Iceland	1991
Krafla	654/2,145	Iceland	1984
Grimsvötn	1,725/5,658	Iceland	1996
Eldfell/Helgafell	215/706	Iceland	1973
Surtsey	174/570	Iceland	1967
Caribbean			
La Grande Soufrière	1,467/4,813	Basse-Terre, Guadeloupe	1977
Pelée	1,397/4,584	Martinique	1932
La Soufrière St. Vincent	1,234/4,048	St. Vincent and the Grenadines	1979
Soufriere Hills/ Chances Peak	968/3,176	Montserrat	1999

Index

[Note: page numbers in *italics* refer to illustrations.]